VOLUME ONE HUNDRED AND EIGHTY FOUR

Progress in
**MOLECULAR BIOLOGY
AND TRANSLATIONAL
SCIENCE**

Advances in Aggregation Induced
Emission Materials in Biosensing
and Imaging for Biomedical
Applications - Part A

VOLUME ONE HUNDRED AND EIGHTY FOUR

PROGRESS IN
MOLECULAR BIOLOGY AND TRANSLATIONAL SCIENCE

Advances in Aggregation Induced Emission Materials in Biosensing and Imaging for Biomedical Applications - Part A

Edited by

RAJESH S. BHOSALE
Department of Chemistry, School of Science, Indrashil University, Rajpur, Mehsana, Gujarat, India

VIJAI SINGH
Department of Biosciences, School of Science, Indrashil University, Rajpur, Mehsana, Gujarat, India

Academic Press is an imprint of Elsevier
50 Hampshire Street, 5th Floor, Cambridge, MA 02139, United States
525 B Street, Suite 1650, San Diego, CA 92101, United States
The Boulevard, Langford Lane, Kidlington, Oxford OX5 1GB, United Kingdom
125 London Wall, London EC2Y 5AS, United Kingdom

First edition 2021

Copyright © 2021 Elsevier Inc. All rights reserved.

No part of this publication may be reproduced or transmitted in any form or by any means, electronic or mechanical, including photocopying, recording, or any information storage and retrieval system, without permission in writing from the publisher. Details on how to seek permission, further information about the Publisher's permissions policies and our arrangements with organizations such as the Copyright Clearance Center and the Copyright Licensing Agency, can be found at our website: www.elsevier.com/permissions.

This book and the individual contributions contained in it are protected under copyright by the Publisher (other than as may be noted herein).

Notices
Knowledge and best practice in this field are constantly changing. As new research and experience broaden our understanding, changes in research methods, professional practices, or medical treatment may become necessary.

Practitioners and researchers must always rely on their own experience and knowledge in evaluating and using any information, methods, compounds, or experiments described herein. In using such information or methods they should be mindful of their own safety and the safety of others, including parties for whom they have a professional responsibility.

To the fullest extent of the law, neither the Publisher nor the authors, contributors, or editors, assume any liability for any injury and/or damage to persons or property as a matter of products liability, negligence or otherwise, or from any use or operation of any methods, products, instructions, or ideas contained in the material herein.

ISBN: 978-0-323-90739-2
ISSN: 1877-1173

For information on all Academic Press publications
visit our website at https://www.elsevier.com/books-and-journals

Publisher: Zoe Kruze
Acquisitions Editor: Ashlie M. Jackman
Developmental Editor: Jhon Michael Peñano
Production Project Manager: James Selvam
Cover Designer: Matthew Limbert

Typeset by STRAIVE, India

Contents

Contributors ix
Preface xiii

1. **Introduction to aggregation induced emission (AIE) materials** 1
   Sujoy Bandyopadhyay, Suresh K. Kalangi, Vijai Singh, and Rajesh S. Bhosale

   1. Introduction 2
   2. Conceptual aspects ACQ and AIE 3
   3. Design principal of AIE molecules 5
   4. AIE active molecules for biomedical applications 6
   5. Conclusion and future prospects 8
   References 8

2. **Aggregation induced emission (AIE) molecules for measurement of intracellular temperature, pH, and viscosity sensing** 11
   Geeta A. Zalmi and Sheshanath V. Bhosale

   1. Introduction 11
   2. Application of AIE molecules for intracellular temperature sensing 13
   3. Application of AIE active molecules for pH sensing 26
   4. AIE active molecules for viscosity sensing 48
   5. Conclusion 58
   Acknowledgments 58
   References 58

3. **Advances in aggregation induced emission (AIE) materials in biosensing and imaging of bacteria** 61
   Mulaka Maruthi and Suresh K. Kalangi

   1. Introduction 62
   2. Aggregation induced materials 64
   3. Photodynamic inactivation 65
   4. Natural biocompatible AIE materials 66
   5. Applications in biosensing 67
   6. Potential application of AIEgens in imaging and killing of bacteria 69
   7. Toxicology aspects of AIEgens 74

8. Future perspectives 75
Acknowledgments 75
References 75

4. **Aggregation-induced emission materials for cell membrane imaging** 81
   Dipratn G. Khandare

   1. Introduction 82
   2. Difficulties in using traditional fluorescent sensors 83
   3. Aggregation-induced emission (AIE) phenomenon 84
   4. Mechanism of aggregation induced emission (AIE) effect 85
   5. Examples of cell membrane imaging 86
   6. Conclusion 94
   7. Future scope 94
   References 95

5. **Aggregation-induced emission luminogens for lipid droplet imaging** 101
   A.H.M. Mohsinul Reza, Yabin Zhou, Jianguang Qin, and Youhong Tang

   1. Introduction 102
   2. Significance of lipid drops study in different organisms 103
   3. Biosynthesis of lipid drops 106
   4. Advancement and challenges in lipid research 108
   5. Recent progress of lipid-specific probes with aggregation-induced emission 113
   6. Conclusions 138
   7. Future remarks 138
   Acknowledgments 138
   References 138

6. **AIE materials for lysosome imaging** 145
   Ankush Gupta, Manoj Kumar, and Vandana Bhalla

   1. Introduction 145
   2. Lysosome viscosity tracing probes 148
   3. Lysosome imaging for detection of other analytes 163
   4. Conclusions 173
   Acknowledgment 175
   References 175

7. Aggregation induced emission (AIE) materials for mitochondria imaging   179

Satish Deshmukh, Madan R. Biradar, Kiran Kharat, and
Sidhanath Vishwanath Bhosale

1. Introduction	180
2. Tetraphenylethene based AIEgens	181
3. TPA derivatives for the mitochondria detecting	192
4. Miscellaneous AIEgens probe for mitochondria imaging	194
5. Future prospectus	200
Acknowledgments	201
References	201

8. AIE materials for nucleus imaging   205

Ankit Singh, Dhara Chaudhary, Aishwarya P. Waghchoure, Ravi N. Kalariya, and Rajesh S. Bhosale

1. Introduction	205
2. AIEgens for nucleus imaging	208
3. Conclusion and future prospects	216
Acknowledgments	216
References	216

9. Aggregation induced emission molecules for detection of nucleic acids   219

Rupesh Maurya, Gargi Bhattacharjee, Nisarg Gohil, Khalid J. Alzahrani, and Vijai Singh

1. Introduction	219
2. Mechanism of action of AIE molecules for nucleic acids detection	
AIE for DNA detection	221
3. AIE for RNA detection	224
4. Conclusion and future perspectives	224
Acknowledgment	224
References	225

Index   *229*

Contributors

Khalid J. Alzahrani
Department of Clinical Laboratories Sciences, College of Applied Medical Sciences, Taif University, Taif, Saudi Arabia

Sujoy Bandyopadhyay
Department of Chemistry, School of Science, Indrashil University, Mehsana, India

Vandana Bhalla
Department of Chemistry, UGC Sponsored Centre for Advanced Studies-II, Guru Nanak Dev University, Amritsar, Punjab, India

Gargi Bhattacharjee
Department of Biosciences, School of Science, Indrashil University, Mehsana, India

Rajesh S. Bhosale
Department of Chemistry, School of Science, Indrashil University, Mehsana, India

Sheshanath V. Bhosale
School of Chemical Sciences, Goa University, Taleigao Plateau, Goa, India

Sidhanath Vishwanath Bhosale
Polymers and Functional Materials Division, CSIR-Indian Institute of Chemical Technology, Hyderabad, Telangana; Academy of Scientific and Innovative Research (AcSIR), Ghaziabad, Uttar Pradesh, India

Madan R. Biradar
Polymers and Functional Materials Division, CSIR-Indian Institute of Chemical Technology, Hyderabad, Telangana; Academy of Scientific and Innovative Research (AcSIR), Ghaziabad, Uttar Pradesh, India

Dhara Chaudhary
Department of Chemistry, School of Science, Indrashil University, Mehsana, India

Satish Deshmukh
Department of Chemistry, MSPMs' Deogiri College, Aurangabad, India

Nisarg Gohil
Department of Biosciences, School of Science, Indrashil University, Mehsana, India

Ankush Gupta
Department of Chemistry, DAV University, Jalandhar, Punjab, India

Suresh K. Kalangi
Amity Stem cell Institute, Amity Medical School, Amity University Haryana, Amity Education Valley Pachgaon, Gurugram, India

Ravi N. Kalariya
Department of Chemistry, School of Science, Indrashil University, Mehsana, India

Dipratn G. Khandare
Department of Chemistry, Modern College of Arts, Science and Commerce Ganeshkhind, Pune, India

Kiran Kharat
KETs V.G. Vaze College, Mumbai, India

Manoj Kumar
Department of Chemistry, UGC Sponsored Centre for Advanced Studies-II, Guru Nanak Dev University, Amritsar, Punjab, India

Mulaka Maruthi
Department of Biochemistry, Central University of Haryana, Mahendergarh, India

Rupesh Maurya
Department of Biosciences, School of Science, Indrashil University, Mehsana, India

Jianguang Qin
College of Science and Engineering, Flinders University, Adelaide, SA, Australia

A.H.M. Mohsinul Reza
College of Science and Engineering; Institute for NanoScale Science and Technology, College of Science and Engineering, Flinders University, Adelaide, SA, Australia

Ankit Singh
Department of Chemistry, School of Science, Indrashil University, Mehsana, India

Vijai Singh
Department of Biosciences, School of Science, Indrashil University, Mehsana, India

Youhong Tang
College of Science and Engineering; Institute for NanoScale Science and Technology, College of Science and Engineering, Flinders University, Adelaide, SA, Australia

Aishwarya P. Waghchoure
Department of Chemistry, School of Science, Indrashil University, Mehsana, India

Geeta A. Zalmi
School of Chemical Sciences, Goa University, Taleigao Plateau, Goa, India

Yabin Zhou
College of Science and Engineering, Flinders University, Adelaide, SA, Australia

Preface

Light is life for all the living beings on Earth. It is hard to point out when exactly mankind began to understand the properties of light and its mechanisms. Light emission is a fascinating photophysical phenomenon, and its different forms have gained the attention of various disciplines of natural sciences for centuries. In the modern era of our scientific generation, short-lived fluorescent light and its long-lived counterpart phosphorescent light have been employed in several chemosensing, biosensing, and bioimaging applications. The traditional organic/inorganic light-emitting materials exhibit excellent light emission in a solution state, but show light emission quenching in aggregates and the solid states. This limits the applications of traditional light-emitting materials in a solid form. Several attempts have been made to address the so-called aggregation caused quenching (ACQ) phenomenon of light emitting small molecules to polymer materials. In 1998, Timothy Swager at Massachusetts Institute of Technology (MIT) in the United States introduced an elegant approach to suppress the molecular aggregation of emissive polymer chains by incorporation of a three-dimensional (3-D) rigid iptycene scaffold, wherein resultant porous polymer exhibits high emission at solid state. In 2000, Ben Zhong Tang introduced the concept of aggregation-induced emission (AIE) for organic materials, wherein 3-D fluorogen illustrates weak emission in the solution state and high emission in the aggregate/solid states. Ever since, tremendous advancements have occurred in the design and development of novel AIE active molecules. Owing to its fundamental and practical importance, researchers worldwide, including material chemists, photophysicists, and biologists, have successively utilized AIE materials in aggregate/solid states for various advanced applications such as chemo/biosensing, imaging, biomedical, organic electronics, and many more.

This volume is especially dedicated to the study of AIE active small molecules to polymer materials, advances in AIE in biosensing, and imaging for biomedical applications. It presents efforts addressing advances in AIE materials and their insightful biomedical applications. Chapter 1 reveals the evolution of ACQ and AIE fundamental concepts. It also deals with the design and transformation of ACQ materials into AIE materials. Chapter 2 reveals the quantification of intracellular pH, viscosity, and temperature by employing AIE active small and polymeric materials. Chapter 3 deals with recent advancements in AIE materials in biosensing and imaging of bacteria. Chapters 4 and 5 cover the last two decades of significant applications of AIE luminogens for cell membrane and lipid droplet imaging. Chapters 6 and 7 give special emphasis to AIE material-based lysosome and mitochondria imaging. Chapters 8 and 9 specifically examine the nucleus and nucleic acid imaging by AIE active molecules. This volume in addition includes fundamental aspects of AIE materials and their advanced biomedical applications; each chapter provides a brief introduction, future prospects, and recent references.

We thank the Elsevier editorial team who helped us to create this volume. We would also like to express our sincere gratitude to peer reviewers who reviewed the chapters with great zeal and suggested valuable corrections. This volume will help future investigators, researchers, students, and stakeholders to perform their research with greater ease. The book will be an excellent basis from which scientific knowledge can grow and widen in the fields of AIE and their applications in different areas.

RAJESH S. BHOSALE
VIJAI SINGH

CHAPTER ONE

Introduction to aggregation induced emission (AIE) materials

Sujoy Bandyopadhyay[a], Suresh K. Kalangi[b], Vijai Singh[c], and Rajesh S. Bhosale[a,*]

[a]Department of Chemistry, School of Science, Indrashil University, Mehsana, India
[b]Amity Stem cell Institute, Amity Medical School, Amity University Haryana, Amity Education Valley Pachgaon, Gurugram, India
[c]Department of Biosciences, School of Science, Indrashil University, Mehsana, India
*Corresponding author: e-mail address: rajeshbhosale24@gmail.com

Contents

1. Introduction 2
2. Conceptual aspects ACQ and AIE 3
3. Design principal of AIE molecules 5
4. AIE active molecules for biomedical applications 6
5. Conclusion and future prospects 8
References 8

Abstract

Idea of introducing aggregation-induced emission (AIE) fundamentally altered the scientific community's perception of classical photophysical processes. Many exciting new possibilities have been coming into light due to the emergence of AIE, such as ability of rapid detection and in analyzing variety of bioactive substances required to monitor the complexed biological processes. This also became a handy tool in elucidating the essential physiological and pathological behaviors of organisms. AIE luminogens (AIEgens) are luminous substances that are either weakly or non-emissive in organic solvents or hydrophobic environment alone, but it gives strong emissive when aggregated along with transforming polarities upon aggregation. Owing to the their outstanding advantages such as rapid turn on/off of high brightness emission, big Stokes shift, excellent photostability, and strong biocompatibility AIEgens have become first choice among bio-inspired probes in biomedicine. In the view of providing basic information on AIE, this chapter give a brief overview of aggregation-caused quenching (ACQ) phenomenon, approaches to transform ACQ to AIE phenomenon, photo-physics of AIE phenomenon, followed by known and reportedly novel AIE active molecules and their biomedical applications.

1. Introduction

Owing to the fact of organic evolution, "Light is the life". It is hard to point out when exactly mankind started understanding the properties of light and its mechanisms, it is widely accepted truth that, Sun is one among the gods of almost all civilizations across the world, may be for the reason of emitting light only. Being a conscious and social animal human might have observed other objectives which are emitting light, including stars to visible fire fly or jelly fish in see without the knowledge of differences in mechanisms emission of light from them. In the modern era of scientific generation fluorescent molecules (probes) provide the advantages including detection of various biomedical entities in in-vitro and in-vivo. Furthermore, fluorescent probes are inexpensive and easy to handle in the form of one-use paper probes. In the past few decades, there has been a lot of attention and much work done into the creation of chromogenic and fluorescent chemical sensors with various biomolecules, biocatalysts, and ions as the key or analytes covering the entire sensor spectrum.[1,2] The interactions between the molecular building blocks of molecular materials are far weaker than interactions between the atomic or ionic units that make up materials such as metals, metal oxides, and semiconductors. While these intermolecular interactions only have a slight effect on the molecular electronic states, they can cause substantial variations in the optical responses when the molecules are separated from each other. The conventional approach of describing the effect of aggregation on electronic absorption and emission spectral features in terms of the relative orientation of the interacting molecules and thus their transition dipoles is frequently used. According to the exciton coupling model, aligning the transition dipoles side by side (H-aggregate) leads to a blue shift of the absorption and reduced emission intensity, while aligning them head to tail (J-aggregate) results in a red shift of the absorption and enhanced emission intensity (Fig. 1).[3,4]

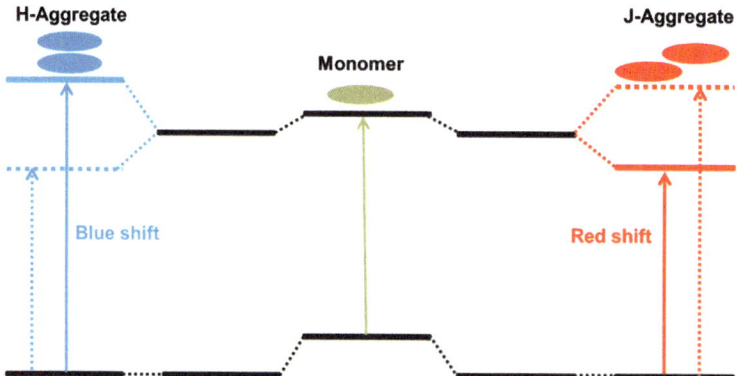

Fig. 1 Energy profile diagram of H and J aggregates.

2. Conceptual aspects ACQ and AIE

Traditional chemo-luminescent agents are earlier version of the modern fluorescent compounds, which showed aggregation-caused quenching (ACQ) behavior. ACQ luminogens, for example, such as pyrene, accumulate structurally due to their π associations, at a high concentration and non-dissolved state or condition. Förster suggested that organic dyes may provide a quenching effect, based on the photoluminescence investigation with the strong π-π stacking interactions at a concentration state (Fig. 2).[5] In addition to collisional interactions among excited molecules and ground-state molecules, the π-π-induced quenching of fluorescence leads to excimers and exciplexes being formed through non-radiative deactivation.[6]

Luminophores have been used to filter for biological molecules in body buffer solutions and track ions in river water. While polar functional groups may be added to a chromophoric unit to yield a hydrophilic entity, unmodified π-conjugated aromatic rings were more likely to accumulate in aqueous media. Due to the fact that there is no solvent, the luminophore concentration approaches its peak in the solid state and the ACQ effect hence becomes most severe.[7] There have been several attempts to address the ACQ dilemma in practical studies of small molecules and polymers. In the late 90's several attempts has been reported to tackle ACQ phenomenon of emissive polymers and small molecules. The most common approach to prevent ACQ was introduced large flexible side-chains. Supramolecular and macromolecular approaches have also been effectively used to diminish molecular aggregation of polymer chains and achieve molecular level separation, and retain emission of polymers.[8] In 1998, Swager and coworkers introduced an elegant approach to suppress the molecular aggregation of emissive polymer chains by incorporating three-dimensional (3-D) rigid shape persistent iptycene scaffold. The incorporation of 3-D iptycene scaffold in polymer backbone prevent the emission quenching in solution as well as film states and create a porous morphology which behaves as a "sponge" structure and served

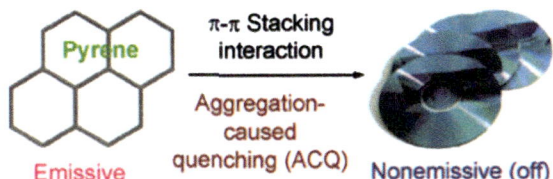

Fig. 2 Strong π–π stacking interactions cause pyrene to agglomerate and "turn off" light emission.

as platform for multiple chemo/biosensor applications (Fig. 3).[9,10] Thereafter series of emissive polymers bearing three-dimensional functional triptycene or iptycene scaffolds have been reported and utilized for several photo physical as well as biomedical applications.[11]

In 2001, a novel luminogen aggregated mechanism was found by B. Z. Tang and his colleagues, which demonstrated aggregability in a positive way rather than negatively luminosis works in a way that increases light intensity, rather than decreases it.[12] Research group of Tang found that phexaphenylsilol (HPS) and its derivatives are non-emitting in dilute solutions, but emissive in concentrated solutions or cast into solid films (Fig. 4). The mechanism was referred to as "aggregation-induced emissions" (AIE) since the light emissions were caused by aggregate creation.[13] AIE mechanism acts just the opposite to the ACQ effect, which makes it possible for technological application instead of simply by "contributing continuously." The AIE method creates a forum for scientists to analyze light effluence from luminogenic aggregates, acquiring knowledge regarding structure–property relationships and insights into operating mechanisms.[14] The tetraphenylethene (TPE) was introduced into the AIE family in 2006, and because of the synthetic simplicity, structural adaptability, and good AIE performance, introduction of TPE into AIE family that triggered a whole new cycle of growth for the engineering.[14] TPE and its diphenylated derivative, for example, were easily produced using simple McMurry reactions of diaryl ketones. In the dilute acetonitrile solutions, hardly unknown fluorescence was found while their nanoaggregates lighted increased emissions by increasing the water percentage. When the water fraction was raised from 0% to 90%, a 1324-fold increase in the fluorescence quantum yields of TPE was observed. Further work indicated that the morphologies of the TPE luminogens influenced their emission. The crystalline aggregates of TPE emitted bluer light than the

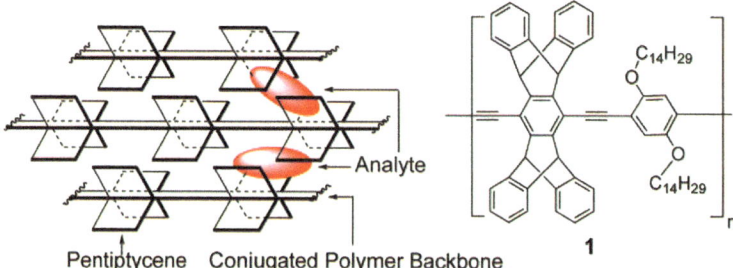

Fig. 3 Polymer-1 bearing three-dimensional rigid iptycene and schematic representation of galleries defined between polymer chains that can host analytes.

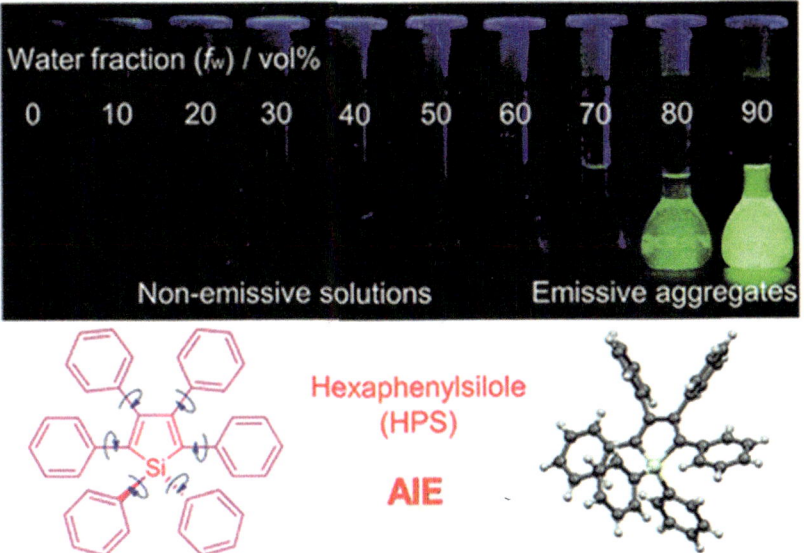

Fig. 4 Fluorescence photographs of solutions and suspensions of hexaphenylsilole (HPS; 20 μM) in THF/water mixtures with different fractions of water (f_w), with molecular and crystal structure of HPS.

amorphous aggregates.[15] Similar effect was also observed in many other AIE systems. Tang *et al.* later created the term crystallization-induced emission (CIE), which refers to the phenomena of AIEgens emitting faintly in their amorphous solid form but emitting intensely in their crystalline phase.[16] Luminogens discovered a diverse variety of innovative uses since they have a morphology-dependent characteristic.

3. Design principal of AIE molecules

In past two decades, development of new AIE systems has been growing for various applications. AIE systems have infiltrated many areas to study and also AIE variety to be preserved a continuous expansion, different types of molecules. Design methods for obtaining a novel molecular architectures that exhibited AIE phenomena which includes suppressing the efficient-stacking of chromophores, insertion of steric shielding to isolate the fluorophore in molecular level. On the other hand, energy must be efficient to dissipate in the solution via molecular movements. This should be accomplished through the incorporation of rotor elements, such as phenyl or aryl substituents, which are consequently common structural features in AIE

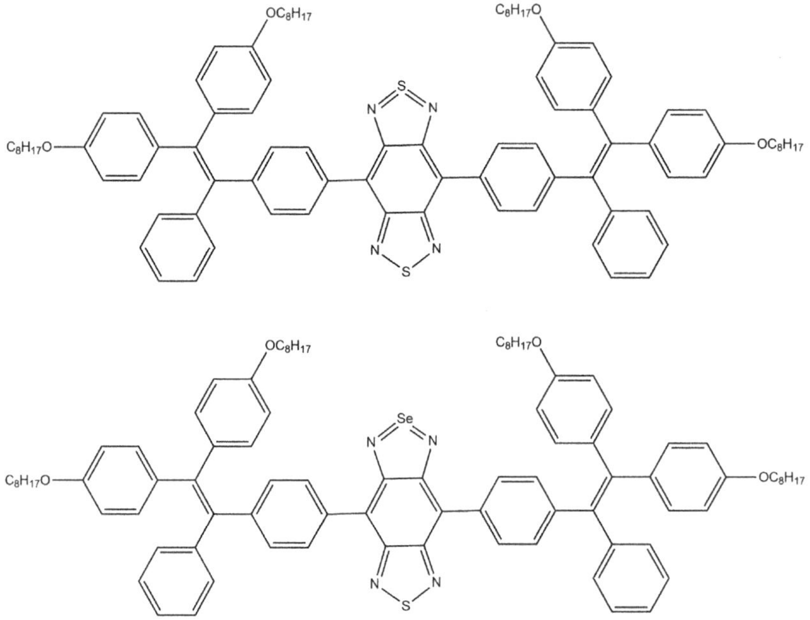

Fig. 5 Molecular structure of NIR active AIE molecules.

active emitters. Fluorogen with simple rotor structures, such as tetraphenylsilols (TPS) and TPE were among the first instances of classic AIEgens to emerge.[17] Likewise incorporation of AIE building blocks to their perimeter, ACQ molecules may become AIEgens.[18,19] Furthermore, introducing donor and acceptor groups into π-conjugation systems of AIEgens allows for more precise tuning of the absorption/emission wavelengths that include the whole visible and NIR spectrums.[20] The recent emergence of novel molecular design principles has been created an environment conducive to understand and inventing unique AIEgen characteristics (Fig. 5).[21]

4. AIE active molecules for biomedical applications

The discovery of AIE has garnered enormous research interests and inspired scientific community a broad range of applications, including AIE based chemical sensors, optoelectronic devices, and a numerous biological probes for biomedical research applications.[22,23] Among all in particular, advancement of AIE in the area of biomedicine has shown an inexhaustible driving force, with enormous promise for resolving a variety of human healthcare problems. Rendering to existing reports, first biological use of AIE was described in 2004, in which the scientists have successfully employed silole-based AIEgens to substantially improve the sensitivity of immunoassays to detect cancer.[24] Following that, the notable characteristics of AIEgens have harvested a great deal of interest, opening the new door by enabling the applicability of AIEgens in a variety of different biological systems. In particular, last decade is the golden era of AIE technique penetrated into research objectives, which are evolving from basic chemical systems to more complex cell and animal models. The use of highly distributed AIEgens and AIE-active luminogens were used in the early stages of research to detect and interacted with a variety of molecules including ions, small molecules, and functional bio-macromolecules (e.g., DNA, proteins, and polysaccharides) (Fig. 6).[25,26] Due to the rapid turning on/off mechanism of, AIEgens were also used for cell imaging and bio-sensing. The tremendous advantages of "all-in-one" nanomaterials have inspired the development of a variety of AIEgens based bio-probes that are not only utilized for in-vitro cell imaging (e.g., cancer cell targeting and bacterium discriminating), but were also scaled up to animal levels for in-vivo applications (e.g., tumor labelling and vascular imaging) (Fig. 6).[27] The emission spectra of AIE probes are being pushed to the limit in order to achieve deep tissue penetration by replacing the nano classical dyes or nanomaterials. Non-emissive nature being in organic solvent and showing self-assembly enabled emission in transformative polarity environments, and highly biocompatibility nature of interacting with biomolecules, made AIEgens are targetable bio-imaging of specific subcellular level to single bio molecular structure. As selectively labelling is high priority in identifying and discriminating the healthy Vs Pathological cell or tissue, AIE technology paved a grand road to enhance the vicinity of biomedical research applications.

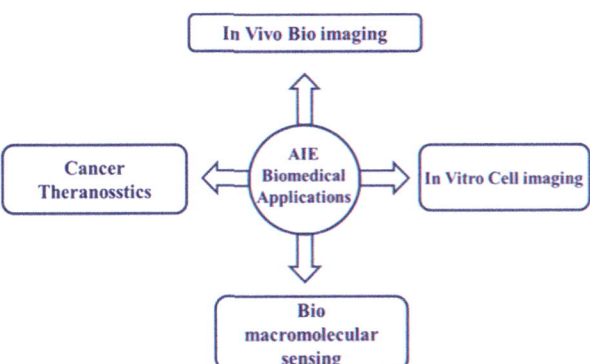

Fig. 6 Schematic representation of AIE molecules used for various biomedical applications.

5. Conclusion and future prospects

AIE based probes named as AIEgens plays crucial roles in biomedical field, which helps in diagnosis of various disease and bio-imaging. The current design of AIEgen probes are based on various aspects including transformation of original ACQ core structure to AIE active structure with good photostability and sensitivity. In the last decade, AIEgens were successfully employed for various biomedical applications such as sensing of intracellular pH, temperature and viscosity, imaging of cell membrane, cytosol, lysosome, mitochondria, nucleus, DNA, RNA, protein fibrils, cancer, photodynamic therapy and even drug delivery, bacteria, fungi, virus, and many more. AIEgens can be further expanded for more advancement with other light emitting features such aggregation induced delayed fluorescence, aggregation induced phosphorescence, aggregation induced circular polarized light emitting for comprehensive biomedical applications.

References

1. Boens N, Leen V, Dehaen W. Fluorescent indicators based on BODIPY. *Chem Soc Rev*. 2012;41:1130–1172.
2. Wu D, Sedgwick AC, Gunnlaugsson T, Akkaya EU, Yoon J, James TD. Fluorescent chemosensors: the past, present and future. *Chem Soc Rev*. 2017;46:7105–7123.
3. Brixner T, Hildner R, Köhler J, Lambert C, Würthner F. Exciton transport in molecular aggregates—from natural antennas to synthetic chromophore systems. *Adv Energy Mater*. 2017;7:1700236.
4. Patra A, Chandaluri CG, Radhakrishnan TP. Optical materials based on molecular nanoparticles. *Nanoscale*. 2012;4:343–359.
5. Hong Y, Lam JWY, Tang BZ. Aggregation-induced emission: phenomenon, mechanism and applications. *Chem Commun*. 2009;29:4332–4353.
6. Hoche J, Schmitt H-C, Humeniuk A, Fischer I, Mitrić R, Röhr MIS. The mechanism of excimer formation: an experimental and theoretical study on the pyrene dimer. *Phys Chem Chem Phys*. 2017;19:25002–25015.
7. Guo R, Li T, Shi S. Aggregation-induced emission enhancement of carbon quantum dots and applications in light emitting devices. *J Mater Chem C*. 2019;7:5148–5154.
8. Frampton MJ, Anderson HL. Insulated molecular wires. *Angew Chem Int Ed*. 2007;46:1028–1064.
9. Yang JS, Swager TM. Porous shape persistent fluorescent polymer films: an approach to TNT sensory materials. *J Am Chem Soc*. 1998;120:5321–5322.
10. Yang JS, Swager TM. Fluorescent porous polymer films as TNT chemosensors: electronic and structural effects. *J Am Chem Soc*. 1998;120:11864–11873.
11. Swager TM. Advanced polymer design and synthesis. *Acc Chem Res*. 2008;41:1181–1189.
12. Hong Y, Lam JWY, Tang BZ. Aggregation-induced emission. *Chem Soc Rev*. 2011;40:5361–5388.
13. Mei J, Hong Y, Lam JWY, Qin A, Tang Y, Tang BZ. Aggregation-induced emission: the whole is more brilliant than the parts. *Adv Mater*. 2014;26:5429–5479.

14. Chen J, Law CCW, Lam JWY, et al. Synthesis, light emission, nanoaggregation, and restricted intramolecular rotation of 1,1-substituted 2,3,4,5-tetraphenylsiloles. *Chem Mater.* 2003;15:1535–1546.
15. Chen Y, Lam JWY, Kwok RTK, Liu B, Tang BZ. Aggregation-induced emission: fundamental understanding and future developments. *Mater Horiz.* 2019;6:428–433.
16. Tong J, Wang YJ, Wang Z, Sun JZ, Tang BZ. Crystallization-induced emission enhancement of a simple stolane-based mesogenic luminogen. *J Phys Chem C.* 2015;119:21875–21881.
17. Cai X, Liu B. Aggregation-induced emission: recent advances in materials and biomedical applications. *Angew Chem Int Ed.* 2020;59:9868–9886.
18. Mei J, Leung NLC, Kwok RTK, Lam JWY, Tang BZ. Aggregation-induced emission: together we shine, united we soar. *Chem Rev.* 2015;115:11718–11940.
19. Xu W, Wang D, Tang BZ. NIR-II AIEgens: a win-win integration towards bio-applications. *Angew Chem Int Ed.* 2021;60:7476–7487.
20. Bhosale RS, Aljabri M, La DD, Bhosale SV, Jones LA, Bhosale SV. Tetraphenylethene derivatives: a promising class of AIE luminogens synthesis, properties, and applications. In: Tang Y, Tang B, eds. *Principles and Applications of Aggregation Induced Emission.* 1st ed. Cham: Springer; 2019. https://doi.org/10.1007/978-3-319-99037-89.
21. Xu S, Duan Y, Liu B. Precise molecular design for high-performance luminogens with aggregation-induced emission. *Adv Mater.* 2020;32:1903530.
22. Lee YH, Kweon OY, Kim H, Yoo JH, Han SG, Oh JH. Recent advances in organic sensors for health self-monitoring systems. *J Mater Chem C.* 2018;6:8569–8612.
23. Zhu C, Kwok RTK, Lam JWY, Tang BZ. Aggregation-induced emission: a trailblazing journey to the field of biomedicine. *ACS Appl Biol Mater.* 2018;1:1768–1786.
24. Chan CP, Haeussler M, Tang BZ, et al. Silole nanocrystals as novel biolabels. *J Immunol Methods.* 2004;295:111–118.
25. Wang M, Zhang D, Zhang G, Tang Y, Wang S, Zhu D. Fluorescence turn-on detection of DNA and label-free fluorescence nuclease assay based on the aggregation-induced emission of silole. *Anal Chem.* 2008;80:6443–6448.
26. Hong Y, Häußler M, Lam JWY, et al. Label-free fluorescent probing of G-quadruplex formation and real-time monitoring of DNA folding by a quaternized tetraphenylethene salt with aggregation-induced emission characteristics. *Chem Eur J.* 2008;14:6428–6437.
27. Lee Y-D, Lim C-K, Singh A, et al. Dye/peroxalate aggregated nanoparticles with enhanced and tunable chemiluminescence for biomedical imaging of hydrogen peroxide. *ACS Nano.* 2012;6:6759–6766.

CHAPTER TWO

Aggregation induced emission (AIE) molecules for measurement of intracellular temperature, pH, and viscosity sensing

Geeta A. Zalmi and Sheshanath V. Bhosale*
School of Chemical Sciences, Goa University, Taleigao Plateau, Goa, India
*Corresponding author: e-mail address: svbhosale@unigoa.ac.in

Contents

1. Introduction	11
2. Application of AIE molecules for intracellular temperature sensing	13
3. Application of AIE active molecules for pH sensing	26
4. AIE active molecules for viscosity sensing	48
5. Conclusion	58
Acknowledgments	58
References	58

Abstract

This book chapter presents insightful growth and progress in the field of sensing especially, temperature, pH, and viscosity sensing. We focus more on aggregation-induced emission (AIE)-active materials for measuring intracellular pH, viscosity, and temperature by means of fluorescence and absorption study. A special emphasis is given on AIE active fluorescent molecules, molecular rotors, polymeric nanomaterials which are considered as the important aspects of sense. It also gives the fundamental and brief understanding between these different AIE active material and its application in biological systems.

1. Introduction

Aggregation-induced emission is an important photophysical phenomenon developed by the Tang group in 2001. In AIE active phenomena a strong light emission takes place from a fluorogen in the solid aggregated state. Initially, the fluorogen in the dilute solution state is non-emissive but emits the intense light very strongly in highly aggregative solid state.[1,2]

Before the development AIE there was another phenomenon developed which is considered detrimental known as aggregation caused quenching (ACQ) in which fluorogens is emissive dilute solution while becomes non-emissive in aggregated state which exactly reverse phenomenon to AIE. Between ACQ and AIE, ACQ is detrimental which led to the discovery of AIE which is the most important and very useful. The discovery of AIE has completely changed and has opened a new way for traditional thinking and has brought up new ideas in designing and synthesizing the novel luminophore.[3-5]

Many fluorophores usually exhibit ACQ (aggregation caused quenching) effect which is caused due to π-π stacking interaction which has deprived application when used in high concentration. AIE another photophysical phenomenon which exactly differs from ACQ where luminophore becomes non-emissive in dilute solution, however, emit strongly in aggregated state. This phenomenon is termed as AIE and the fluorophore exhibiting phenomenon termed as AIEgens. This AIEgens are widely used as an excellent fluorescent candidature for biosensor due to its high sensitivity and selectivity, fast response time, and low background noise which inspired the researcher to design and synthesize luminogens that can help to study temperature, pH, and viscosity sensitive behavior and potentially applicable for cell imaging.

Moreover, the AIE molecules eliminate the excitation by a non-radiative decay process in a dilute solution. As there is the formation of aggregate the molecular rotation is restricted which results in radiative decay and the light emission takes place in the aggregated state. Nowadays these AIE molecules have a widespread application in detecting the various components in the smaller organisms.[6] AIEgens are not only useful in vitro but these fluorogens are utilized for biological imaging of the molecules, cell, tissues, and components of organism.[1] The AIE active molecules are also widely used as optical sensors for the assay of proteins, peptides, and amino acids and can be also utilized for monitoring the conformational changes and widely used for detection of explosive. Moreover, scope and application of AIEgens in the recent progress has extended in the various scientific field.

Nowadays many scientists have increasingly exploited the concept of AIE luminescent material for potential practical application. Due to growing research and efforts by several groups have developed various kinds of stimuli-responsive AIE active materials such as fluorescent organic molecules, nanoparticles, different molecular rotors, polymer-based nanogels these AIE active molecules can be used to study the responses to various environmental changes such as light, temperature,[7] pressure,[8] pH,[9] and viscosity.[4] Among all these parameters temperature, pH and viscosity are the major factors responsible for biological activities occurring in the living system.

2. Application of AIE molecules for intracellular temperature sensing

In recent years molecular sensors have received enormous attention and becoming increasingly attractive due to its wide biomedical application. The change occurring in the fluorescent material are generally characterized by applying external stimuli such as light, temperature, pH. Among all these temperature-sensitive materials shows the phase transition above and below certain temperature due to which it shows a huge potential application in many scientific fields.[10,11]

Temperature is a fundamental physiological parameter in various biological activities and reaction taking place within the living system such as the enzyme activity, gene expression, cell division, and energy metabolism. Abnormal increase in temperature leads the cellular microenvironment to dysfunction which results in different types of cellular disorder such as inflammation and cancerous cell growth.

There are several other conventional methods and devices which can be used to measure temperature but due to small confined space thus not possible to measure the temperature in the individual live cells. Since the fluorescent molecular sensors shows good sensitivity, high spatial resolution and rapid response explored its increasing demand in a biological application. Unfortunately, still it is very challenging for many researcher to study the accurate detection of local temperatures in individual living cell.[12]

Over the past few years, numerous florescent thermosensitive materials have been developed. However, these thermosensitive materials limits its application in various field because of poor hydrophilicity, unwanted leakage from the cell, poor structural stability, and non-negligible biotoxicity. In order to overcome these problems many fluorescent polymeric thermosensitive materials have been designed, synthesized, and applied to determine the intracellular temperature in the living cell because of their outstanding photostability, water stability, and biocompatibility.[13] Mostly fluorescent polymeric thermosensitive material consists of two parts thermo-responsive unit and a polarity sensitive conventional fluorophore unit.

Thermo-responsive elastin-like polypeptide ELP(S)[10] with Aggregation-induced emission conjugate was constructed having thermal responsive property which is extensively used for in vivo applications. ELPs comprised of val-pro-gly-xaa-gly (VPGXG) pentapeptide with amino acids as guest residue except proline. This ELP's undergoes a reversible phase transition at their inverse transition temperature (Tt). Since ELP disperse well at below Tt and appears as disordered structures but above Tt ELPs forms aggregate in water. The Tt of ELPs can be altered by changing the parameters such as change in composition of amino acids and molecular weight and thus Tt can be accurately adjusted between 0 °C and 100 °C. Synthesis of ELPs40-TPE as successful where the TPE-COOH molecule underwent interaction with ELPs40 (Fig. 1). Further, 12% sodium dodecylsulphate polyacrylamide gel electrophoresis was used to study interaction of TPE-COOH to ELPs40-TPE. However, ELPs40-TPE were well studied and characterized by Circular Dichroism spectra, gel electrophoresis, and optical studies of ELPs40-TPE and ELPs40 were done by UV–visible absorption spectra and fluorescence emission spectra.

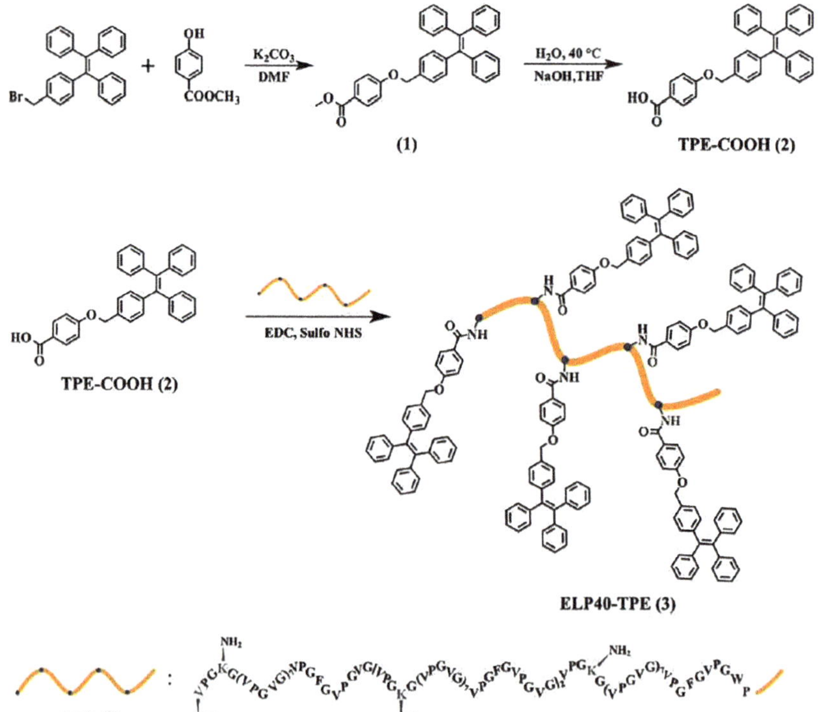

Fig. 1 A schematic illustration of the ELP-40 TPE. *Reprinted from Chen Z, Ding Z, Zhang G, Tian L, Zhang X. Molecules. 2018;23 with permission of MDPI.*

AIE active phenomenon of ELPs40-TPE was studied in DMF/Water in various water fraction. Initially in low water fraction ELPs40-TPE completely dispersed in pure DMF and shows weak fluorescence but upon incremental addition of water fraction to 90% ELPs40-TPE becomes highly fluorescent. The Restricted intramolecular rotation (RIR) and restricted intramolecular vibration (RIV) are two well-known mechanisms responsible for AIE active nature of ELPs-40 molecule.

In an aggregated state there is a restriction of such motion which blocks a non-radiative pathway enforcing AIEgens to exhibit strong fluorescence emission. Thus the random coil conformation, hydrogen bonds in acid amide groups and water molecules makes the environment rigid and ELPs40-TPE found to show hydrophobic nature. Due to rigid nature and hydrophobic effect restricts the rotation and inhibit the non-radiative pathway showing strong fluorescence emission.

The temperature response of the ELPs40-TPE was studied in an aqueous solution and revealed that the fluorescence intensity gradually decreases with increase in temperature by 5 °C in steps at specified temperature from 25 °C to 50 °C. The fluorescence intensity of ELPs40-TPE was found to be in good agreement with correlation coefficient of 0.9953. The reversible response of ELPs-40 TPE was studied at higher temperature by heating the solution to 50 °C and lowering temperature to 25 °C thus found that the fluorescence intensity of ELPs-40 TPE increases gradually when the temperature of solutions returns to 25 °C. However during this temperature change from 25 °C to 50 °C there are two effects observed in which at higher temperature leads this AIEgens to exhibit weaker fluorescence and further leading to enhancement in fluorescence when the intramolecular rotations of TPE moiety gets hindered upon lowering the temperature (Fig. 2).

Saha[13] and co-workers synthesized a well-known thermoresponsive aggregation-induced emission (AIE) active non-conjugated poly(N-vinyl caprolactam) as a fluorescent thermometer for intracellular temperature sensing. This polymeric fluorescent thermometer consists of thermoresponsive moiety and polarity sensitive chromophore unit incorporated by copolymerization method. However, the similar strategy was designed and utilized for construction of PNVCL by free radical polymerization (FRP) with commercially available N-vinyl caprolactam (Fig. 3) the temperature-driven aggregation behavior, AIE characteristics and excellent cytocompatibilty have made these PNVCL molecules as a potential fluorescent polymeric thermometer for intracellular temperature determination. Thus making PNVCL as an excellent biomarker for intracellular temperature measurement in MCF-7 cells. Initially, the cell culture was treated with

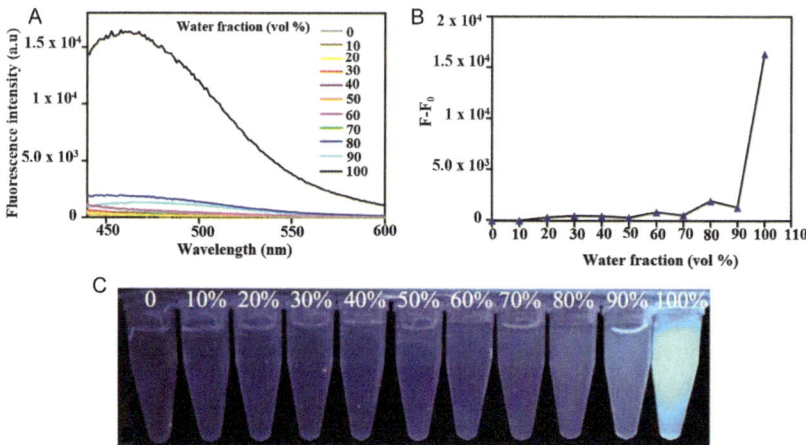

Fig. 2 (A) A fluorescent spectra of ELPs-40-TPE (conc.5 μM, $\lambda_{ex}=320$ nm) in DMF-water. (B) Represents the highest fluorescence intensity of ELPs-40 TPE (conc.1 μM, $\lambda_{ex}=320$ nm) in DMF-water. (C) A fluorescent image of ELPs-40-TPE under UV light at 365 nm in different water fraction. *Reprinted from Chen Z, Ding Z, Zhang G, Tian L, Zhang X. Molecules. 2018;23 with permission of MDPI.*

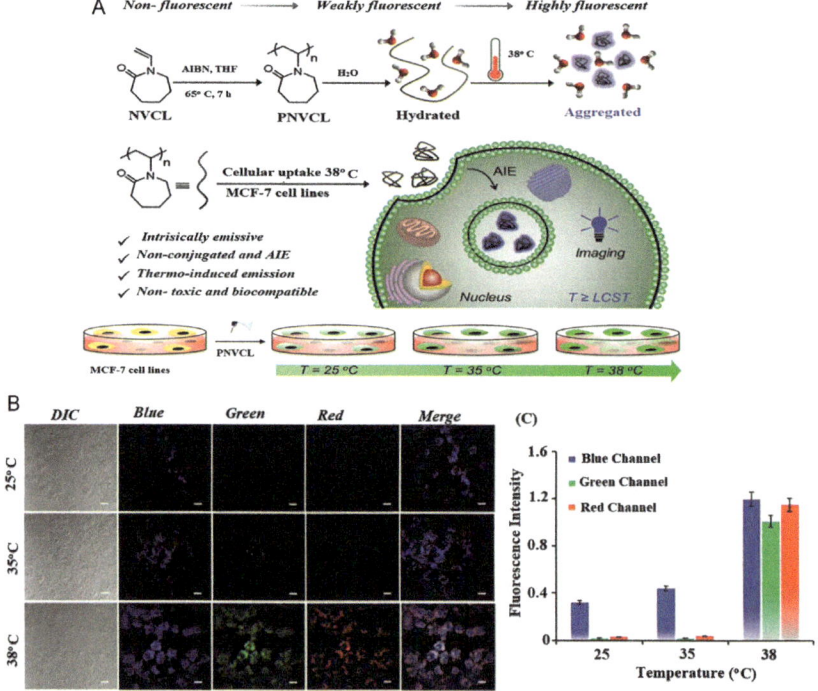

Fig. 3 (A) A schematic demonstration of non-conjugated PNVCL fluorescent thermometer for intracellular temperature sensing, illustration of MCF-7 cells upon incubation with PNVCL fluorescent probe at different temperature. (B) Live cell MCF-7 cells incubated with PNVCL at different temperatures. (C) Graph demonstrates temperature concentrating fluorescence intensities of MCF-7 cell with PNVCL. *Reprinted from Saha B, Ruidas B, Mete S, Mukhopadhyay CD, Bauri K, De P. Chem Sci. 2020;11:141–147 with permission of The Royal Society of Chemistry.*

PNVCL for 24 h at 25 °C, 35 °C, and 38 °C. The confocal images of the cells showed weak blue fluorescence at all these temperatures. However, there was remarkable fluorescence inside cell when incubated above the lower critical solution temperature (LCST), i.e., 38 °C in all three channels. This is found to be in good agreement with the temperature-induced fluorescence enhancement phenomenon. Moreover, a modest increase in fluorescence enhancement observed above 38 °C and there is a significant 50 and 29-fold enhancement in fluorescence with green and red fluorescence as the temperature is changed by 3 °C ranging from 35 °C to 38 °C. Thus implies that the cellular intake of PNVCL fluorescent probe is well studied and controlled by temperature of the cell media.

A polymeric thermo-responsive fluorescent AIE polymers was synthesized by Li et al. through free-radical copolymerization of N-isopropylacrylamide (NIPAM) monomer, oligo (ethylene glycol) methacrylate (OEGMA) monomer or methyl methacrylate (MMA) monomer, and the fluorescent tetraphenylethylene monomer.[14] The hydrophilicity of the copolymer can be altered by changing the composition of NIPAM and OEGMA/MMA monomers (Fig. 4). However, due to hydrogen bonding between the polymer chain and water molecules formed below LCST the polymer exhibits good solubility resulting to weak emission while above LCST polymer shows high emissive property due to interchain and intrachain hydrogen bonds within the polymer.

The temperature dependent transmittance studies for copolymer reveals that in aqueous solution all polymers were transparent with **P1d** but upon heating to 50 °C the solution turns turbid. Since the fluorescence of the AIE active polymer is completely dependent on polymer conformation the temperature-dependent fluorescence studies were performed (Fig. 5A). Thus, photoluminescence of polymer under investigation **P1d** was recorded by heating slowly from 32 °C to 50 °C. It was observed that as the temperature increases the emission peak gradually increases at 480 nm.(Fig. 5B). However, a good linear relationship obtained for **P1d** in the temperature range of 36–42 °C. Similar temperature dependent emission studies were done for other polymers. The repeatability and reversibility for aqueous solution of **P1d** was investigated by heating the solution to 50 °C and by cooling at 27 °C. The fluorescence intensities were analyzed and recorded at each cycle in the temperature range of 43 °C and 33 °C.

Wang et al. described TPE-based AIE active temperature sensitive nanoparticle with aggregation-induced emission effect that can be used for cellular imaging.[15] This thermo-responsive organic nanoparticle with the AIE effect was synthesized by tetraphenylethene derivative as initiator based

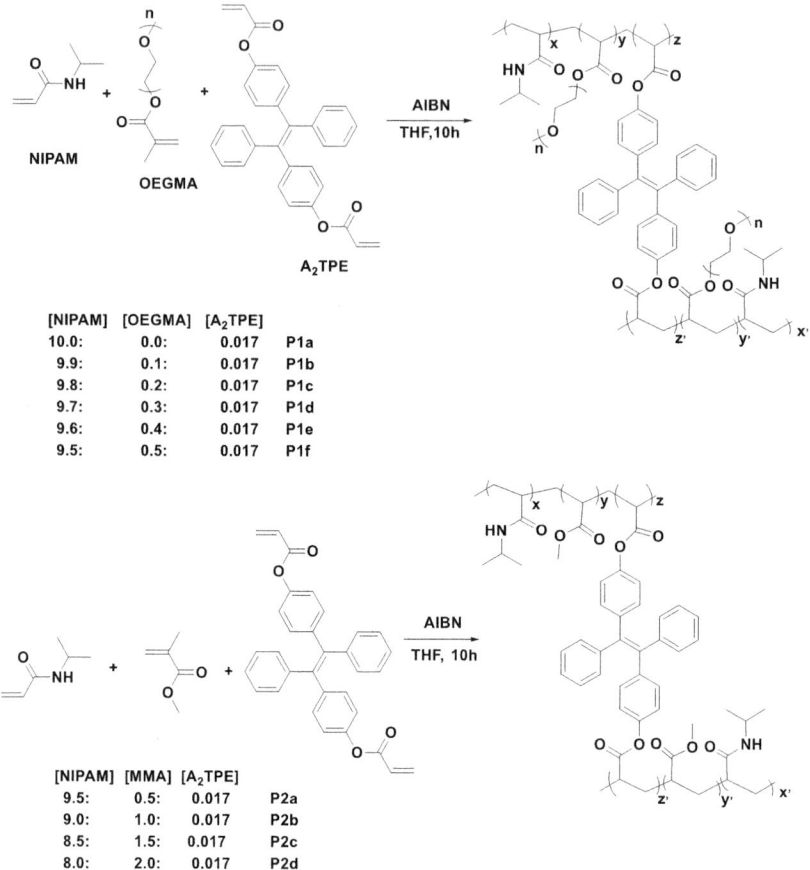

Fig. 4 Synthetic illustration for thermo-responsive AIE active polymers.

poly(N-isopropylacrylamide) (TPE-PNIPAM)[16,17] by atom transfer radical polymerization (ATRP) which displayed good self-assembling nature in to nanoparticle in water. As shown in schematic representation the obtained TPE-PNIPAM exhibiting AIE characteristics with self-assembling into nanoparticle in water (Fig. 6).

The size of **TPE-PNIPAM** and fluorescence property of nanoparticle are highly temperature dependent thus these two factors size and optical properties can be tuned by varying temperatures. The **TPE-PNIPAM** showed good AIE active property. The **TPE-PNIPAM** was studied for self-assembly results revealed that the polymer gets self-assembled in an aqueous solution due to its amphiphilic property. The **TPE-PNIPAM** contains hydrophobic TPE as an aggregated core and **PNIPAM** as

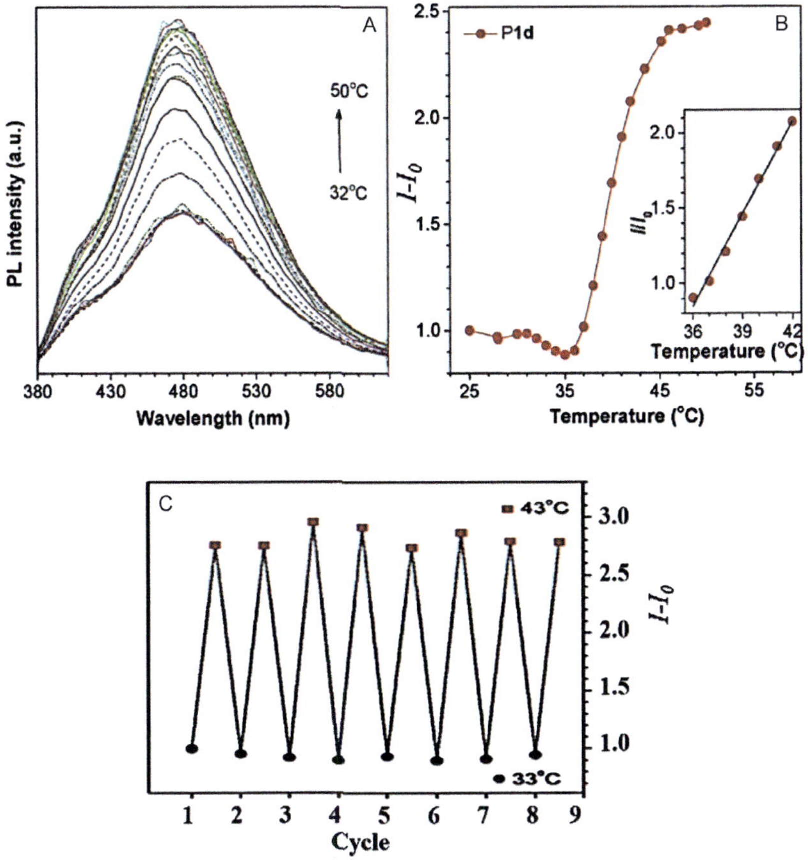

Fig. 5 (A) The fluorescence spectra of **P1d** aqueous solution at various temperature 32–50 °C. (B) A plot of fluorescence intensity vs the temperature for **P1d** at 480 nm. Inset: The linear fitting plot of I/I_o at different temperature ranging from 36 °C to 42 °C. (C) Plot illustrating fluorescence intensities of heating at 43 °C and cooling process at 33 °C at 478 nm. *Reprinted from Li T, He S, Qu J, Wu H, Wu S, Zhao Z, Qin A, Hu R, Tang BZ. J Mater Chem C. 2016;4:2964–2970 with permission of The Royal Society of Chemistry.*

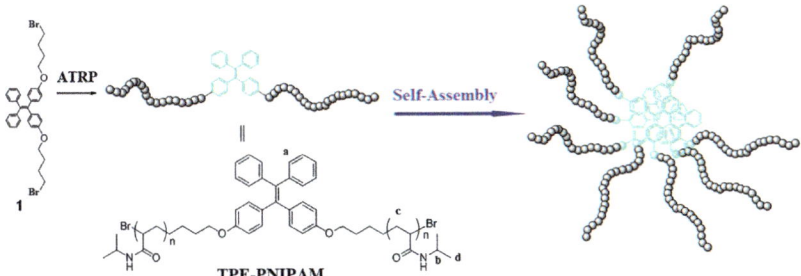

Fig. 6 Schematic presentation of the formed **TPE-PNIPAM** nanoparticle in water. *Reprinted from Wang Z, Yong TY, Wan J, Li ZH, Zhao H, Zhao Y, Gan L, Yang XL, Xu HB, Zhang C. ACS Appl Mater Interfaces. 2015;7:3420–3425.with permission of The American Chemical Society.*

hydrophilic extended chain in water which resulted into formation of nanoparticle 200 nm size. However, the **TPE-PNIPAM** shows good thermoresponsive property which was determined in aqueous media by heating polymer to 50 °C. thus the temperature change results in phase transition from hydrated to dehydrated granules. Initially the solution is transparent at room temperature but when temperature reaches to LCST solution shows turbidity (Fig. 7). Moreover, the fluorescence studies of the **PNIPAM**[18,19] polymer revealed that due to increase in temperature there is decrease in fluorescence intensity due to formation of aggregate nanoparticle which quickens the molecular motions and intramolecular rotations. Thus, the fluorescence intensity decreases with increasing temperature. Similar work was done for **TPE-NIPAM** with P1b and P1c which almost showed similar behavior.[20]

There are several dual and multi responsive polymers synthesized which are sensitive to both temperature as well as for pH. Herein, another AIE active dual mode hydrogel poly(N-isopropyl acrylamide-*co*-tetra(phenyl) ethene acrylate)/poly(methacrylic acid) was synthesized by Zhou and co-workers.[21] This AIE active copolymer was prepared by copolymerization method through N-isopropyl acrylamide (NIPAM) and TPE based tetraphenylethene acrylate as monomer unit (Fig. 8).

Co-polymer from **P1 to P6** showed good AIE active property in THF solvent, AIE activity explained with the help of **P4** polymer initially, the polymer shows non-emissive nature but fluorescence started increasing

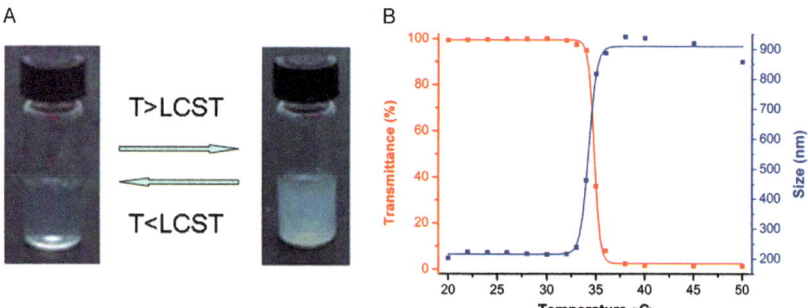

Fig. 7 (A) The images of vial containing the solution TPE-PNIPAM in water at different temperatures (left) at 20 °C and (right) at 50 °C. (B) The plot for TPE-PNIPAM in water against (red line) Transmittance and (blue line) hydrodynamic size vs temperature. *Reprinted from Wang Z, Yong TY, Wan J, Li ZH, Zhao H, Zhao Y, Gan L, Yang XL, Xu HB, Zhang C. ACS Appl Mater Interfaces. 2015;7:3420–3425 with permission of The American Chemical Society.*

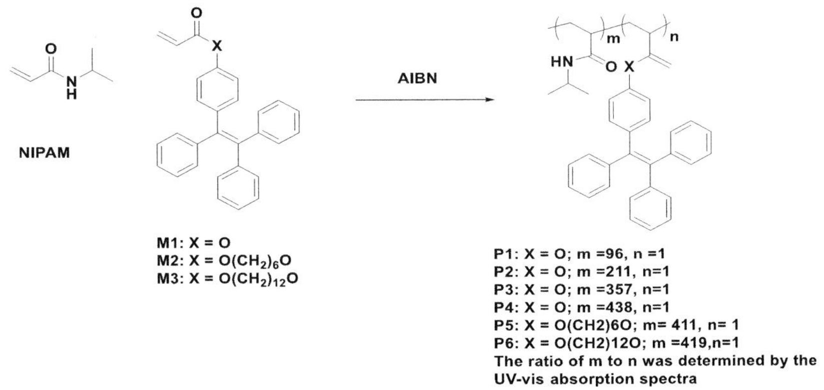
NIPAM
AIBN
M1: X = O
M2: X = O(CH2)6O
M3: X = O(CH2)12O
P1: X = O; m =96, n =1
P2: X = O; m =211, n=1
P3: X = O; m =357, n=1
P4: X = O; m =438, n=1
P5: X = O(CH2)6O; m= 411, n= 1
P6: X = O(CH2)12O; m =419,n=1
The ratio of m to n was determined by the UV-vis absorption spectra

Fig. 8 Schematic illustration describing monomer copolymers (**P1–P6**).

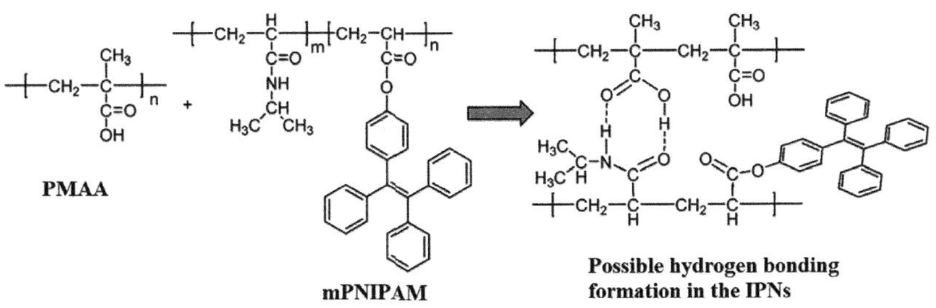
PMAA
mPNIPAM
Possible hydrogen bonding formation in the IPNs

Fig. 9 The systematic schematic illustration between two networks in IPNs and illustration for hydrogen-bond formation between PMAA and PNIPAM. *Reprinted from Zhou H, Liu F, Wang X, Yan H, Song J, Ye Q, Tang BZ, Xu J.* J Mater Chem C. *2015;3:5490–5498 with permission of The Royal Society of Chemistry.*

as the water fraction increases above 70%. It shows that TPE maintains its AIE characteristics even after incorporating in polymer NIPAM. The fluorescence studies for P4-PMAA were done in THF/H_2O at $\lambda_{ex}=318$ nm. The glass transition temperature is higher by 7–9 °C. The complexes are nonemissive in THF but become emissive upon addition of water fraction. However, due to formation of interpenetrating polymer network (IPNs) by interpolymer hydrogen bond interaction the fluorescence gets enhanced as pH values decreases The possible hydrogen bond formation in IPNs as shown in (Fig. 9).

However, the photoluminescence DLS studies shows that the IPNs represents the phase transition behavior upon heating which can be further modified by pH change. This reveals that decrease in pH from 7.0 to 4.0

resulting in to decrease in LCST of IPNs by 5 °C concerning PNIPAM. Hence below the LCST due to change in pH values leads to dissociation of the formed interpolymer hydrogen bond and thus IPNs can fold itself cooperatively and can form the compact structure without any loss in solubility. The temperature-dependent fluorescence studies of IPNs4 displays three stages in fluorescence transitions as a big hump in the emission intensity of IPNs when it was heated in temperature region of 20–27.5 °C and reaches the highest fluorescence emission intensity at 27.5 °C. A stable emission intensity region was recorded in the temperature range from 27 °C to 35 °C illustrating the thermal tolerance. Further reduction in emission intensity due to increase in temperature to 50 °C which thermally activates the molecular motion of TPE unit. The TPE derivative that is connected to the PNIPAM chains shows sensitive response to the phase transition of IPNs. Therefore, this can be utilized as a fluorescent probe to trace IPNs structural changes that takes place due to pH and temperature as shown in Fig. 10, which illustrates that when the pH is (7.0–5.2) > pK_{int} most of the carboxylic acid groups are ionized as very less hydrogen bonding that is formed between PMAA and PNIPAM. The LCST of PNIPAM polymer changes as the composition of PMAA increases due to which the hydrophobic nature of PNIPAM polymer inhibits the complexes that provides high cooperativity of phase transition. Similar trend in fluorescence transition is observed in P4. At pH, less than pK_{int} (4.6, 4.0, 3.4 and 2.8) as the ratio of carboxylic acid groups with PMAA are protonated more number of hydrogens are liberated to form IPNs4 thus affecting the phase transition even in the presence of very low concentration of PMAA. Since the hydrophobic blocks are shorter IPNs4 complex shows low transition cooperativity. Hence IPNs are considered as the convenient source in drug delivery system and the process is further examined by fluorescence spectroscopy.

Meng et al. designed and synthesized twisted intramolecular charge transfer (TICT) based temperature sensitive organic ratiometric fluorescent thermometer which exhibit good AIE active characteristics with donor-π-acceptor luminogen, 2-([1,1′-biphenyl]-4-yl)-3-(4-((E)-4-(diphenylamino) styryl) phenyl) fumaronitrile (TBB) which represents excellent temperature sensitive nature.[22] This thermo-responsive TBB&R110@F127 nanoparticle (TRF NPs) was further modified and prepared by encapsulation of thermosensitive TBB and Rhodamine 110 dye into an amphiphilic polymer matrix F127 for which is considered as fluorescent thermometer for intracellular

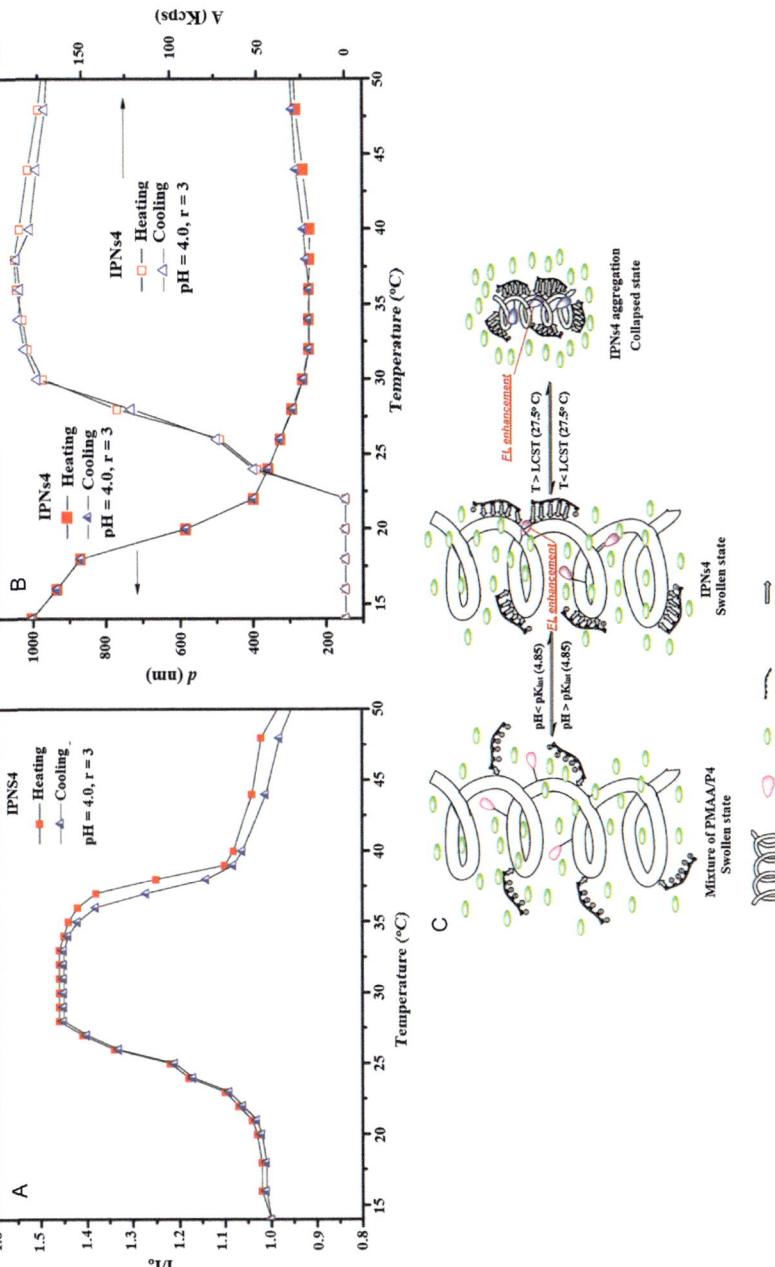

Fig. 10 (A) A plot demonstrates heating and cooling cycle against I/I_0 vs temperature. (B) A graph represents the particle size and turbidity (kcps) vs temperature of IPNs4 with P4 conc.(0.50 mg mL^{-1}), $r = 3.0$, at different pH = 4.0, 10.0 mM with Na$_2$HPO$_4$—citric acid buffer I_0 at 469 nm at $\lambda_{ex} = 318$ nm. (C) Diagrammatic representation at different pH and temperature for P4-PMAA complex. *Reprinted from Zhou H, Liu F, Wang X, Yan H, Song J, Ye Q, Tang BZ, Xu J. J Mater Chem C. 2015;3:5490–5498 with permission of The Royal Society of Chemistry.*

temperature sensing. TRF NPs is highly temperature sensitive and works in the temperature range of 25–65 °C and shows excellent emission reversibility. However, the TRF NPs fluorescent probe can be further used for an intracellular temperature change from 25 °C to 53 °C. The thermoresponsive characteristics of TRF NPs were further investigated at different temperatures.

The fluorescence emission intensity decreases with increasing temperature at 680 nm but as the temperature increases the PL intensity remains unchanged at 520 nm. TRF NPs were further examined for it stability in the temperature range from 25 °C to 65 °C. It was observed that there was no significant change shown for TRF NPs at same temperature thus indicating the good stability (Fig. 11).

Pandey et. al[23] synthesized a fluorescent anthracene based external stimuli responsive based bis-(anthryl)boron fluoride $(B(An)_2F)$ that easily exchange fluorine in $(B(An)_2F)$ with $C_6H_4NH_2$ and $C_6H_4NMe_2$. However, Bis(anthryl)boryl aniline AIEgen can be well used for sensing application. This fluorescent molecular rotors belongs to the donor–acceptor (D-A) class of molecules because of conformation-dependent multiple emission behavior. Depending on temperature the structure of the compound **1** and **2** shows different fluorescence emission color change. (Fig. 12). Due to steric effect both the compounds **1** and **2** show different shifts upon temperature change. In the compound **1** due to less steric demands and weaker donor strength NH_2 the emission maxima does not show any shift with temperature change. But in the case of compound **2** due to the presence of steric crowding with stronger donor NMe_2 group the stokes shift of individual emission with well resolved emission. Time and temperature-dependent emission studies were done in toluene due to low melting −95 °C (178.15 K) and high boiling 110.6 °C (383.75 K).

At lower temperatures due to rigid frozen solvent matrix and due to high viscosity stabilizes the local excited state. There was no significant emission color change observed for compound **2** in temperature range from 80 to 180 K. However, upon increasing temperature, it lowers the viscosity and at 178.15 K the solvent starts melting thus making the medium relatively more fluidic which led to formation of PICT and TICT state that induces the thermo-fluorochromatic changes from cyan to yellow-orange. The temperature range of 80–300 K is considered suitable working range for Bis(anthryl)boryl aniline AIEgen derivative. Various other triaryl pyrene-based molecules were synthesized for sensing temperature.[24,25]

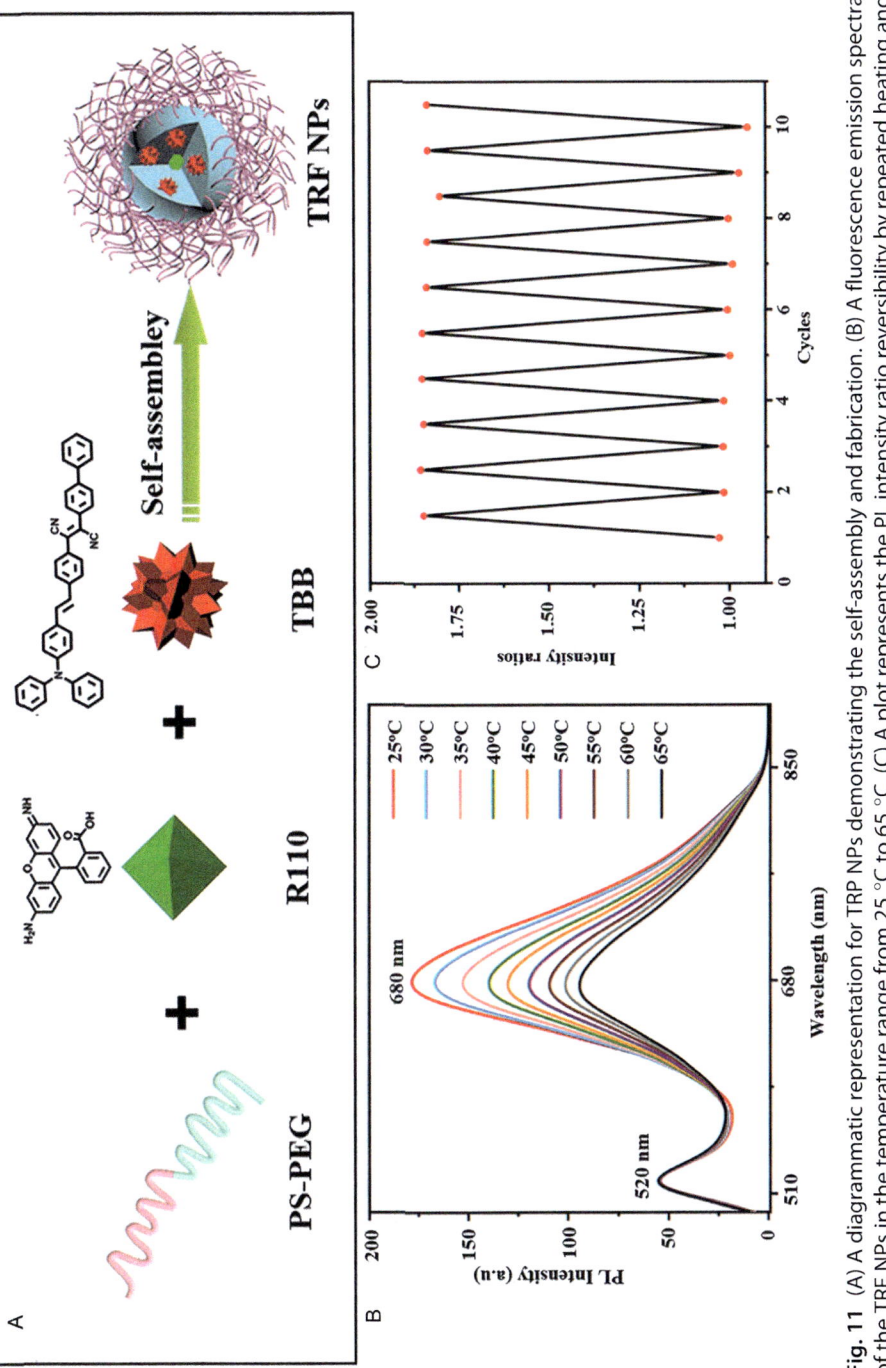

Fig. 11 (A) A diagrammatic representation for TRF NPs demonstrating the self-assembly and fabrication. (B) A fluorescence emission spectra of the TRF NPs in the temperature range from 25 °C to 65 °C. (C) A plot represents the PL intensity ratio reversibility by repeated heating and cooling of TRF NPs. *Reprinted from Meng L, Jiang S, Song M, Yan F, Zhang W, Xu B, Tian W. ACS Appl Mater Interfaces. 2020;12:26842–26851 with permission of American Chemical Society.*

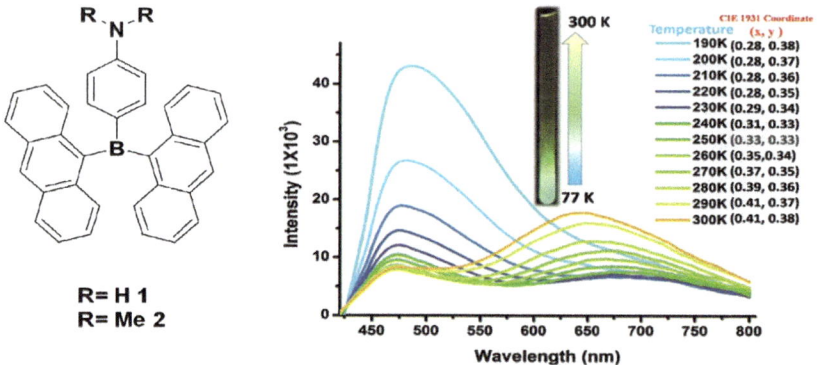

Fig. 12 A structure of the compound **1** and **2** used for sensing viscosity and temperature and the PL spectra of Compound **2** representing the spectral color changes with increasing the temperature from 190 to 300 K. *Reprinted from Pandey UP, Thilagar P. Adv Opt Mater. 2020;8:1–16 with permission of Wiley-VCH.*

3. Application of AIE active molecules for pH sensing

pH plays a very crucial role in terms of environmental, chemical and biological processes therefore accurate pH sensing has become an important factor in the various research field of scientific research and technology. The pH of the body fluid and intracellular fluid in healthy individuals, ranges from 7.35 to 7.45.[26] Generally, regular pH measurement is carried out by using glass-pH electrode which is extensively used as pH sensing device. The glass pH electrode allows pH sensing in wide pH range from acidic to basic to neutral condition due to its low detection limit, interference free with excellent stability and reproducibility. Moreover due to lack of rigid design temperature dependent response and mechanical fragility the glass-pH electrode reduces applicability. Instead, this glass-pH electrode has seriously limited applications in cellular pH sensing.

The difficulties arising from a glass-pH electrode, can be resolved by using optical pH sensing the widely accepted and frequently used being used. Optical pH sensing is used mostly when a glass-pH electrode is not adoptable. Optical pH sensing is well characterized by distinct absorption and fluorescence changes at different pH values that takes place by protonation and deprotonation mechanism. Moreover, higher sensitivity and use of lower concentration of indicators makes the fluorescent indicators more superior than absorption-based pH indicators. This small fluorescent pH indicators can be well employed in sensing intracellular pH sensing in physiological and pathological pathway due to its potential non-invasive nature and real-time imaging. Low solubility and limited response range are some limitations of fluorescent pH indicators in aqueous solution. Usually fluorescent indicators suffer from single fluorescence intensity with pH variation, interference and variation in source intensity along with change in concentration of the indicators. It is noteworthy that most of them undergo significant toxicity, non-specific bindings, and need of other essential sources for transportation in the cell for determining intracellular pH.[27]

Chen and coworkers[28,29] deigned an AIE active TPE-Cy zwitterionic hemicyanine fluorogen comprising of tetraphenylethene (TPE) derivative and N-alkylated indolium moiety with sensing array by switch + knob effect.(Fig. 13). Because of AIE effect and chemical reactivity toward H+/OH the TPE-Cy can sense pH in broad range. TPE-Cy exhibits different emission colors and emission intensity at different pH. It is observed at pH 5–7 shows intense to moderate red emission, at pH 7–10 a noticeable weak to no emission and finally at pH 10–14 no emission to strong blue emission. A good linear regression emission intensity in the physiological pH range from 5 to 7 to acidity which explores the dye as an excellent fluorescent indicator for determining intracellular pH. The chemical reactivity of the probe toward OH-/H+ "(switch)" and AIE effect attribute to "(Knob)" which gives valuable information for designing TPE-Cy fluorescent probe for pH. This report suggests the a pH-sensitive fluorogen can be utilized for determining intracellular pH sensing and tracking. Due to biocompatibility and cell permeability of the TPE-Cy dye shows that after penetration in cells it displays high sensitive response from acidic to basic compartment over the wide range of physiological pH with intense red and blue channels that serves as an indicator for local proton concentration. Thus, confocal microscopy, radiometric analysis and flow cytometry confirms the utility of TPE-Cy in monitoring intracellular pH and cell imaging.

Li and coworkers[30] reported ratiometric AIE active fluorescent 2-(5-(4-carboxyphenyl)-2-hydroxyphenyl) benzothiazole compound. Depending on nature of the solvent the compound showed different fluorescence emission intensity which may be due to presence of different luminescence processes. Multi-fluorescence emission of the compound originated from prototropic equilibria in different solvents. The multi-fluorescence emission characteristics in the compound in dispersed and aggregated state may be due two processes excited state proton transfer and restriction of intramolecular rotation. Therefore the synthesized fluorescent material finds its applicability

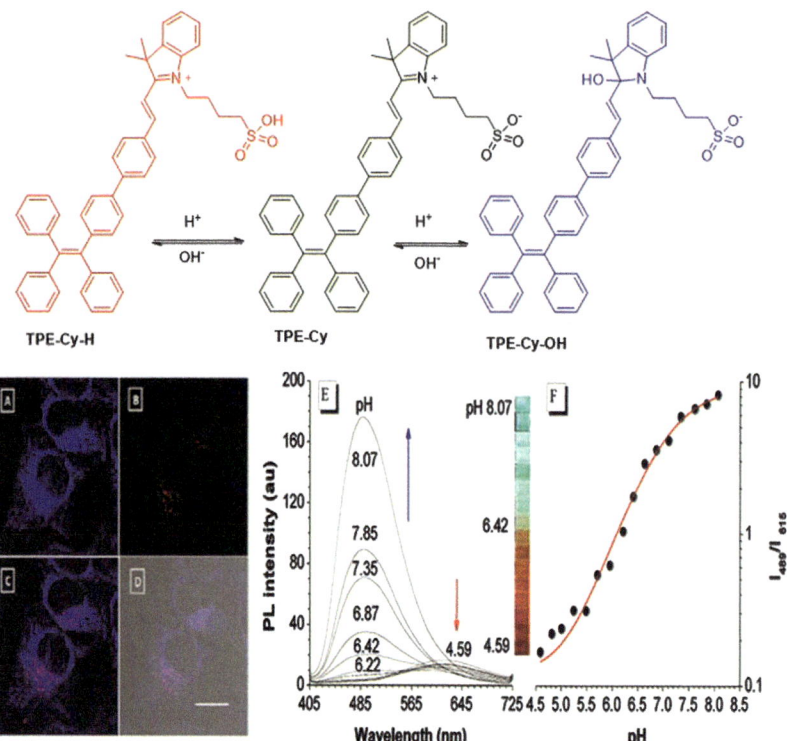

Fig. 13 Schematic illustration showing mechanism of pH change response of **TPE-Cy**. (A) A confocal images of TPE-Cy on Hela cells at 405 nm and (B) confocal image at 488 nm (C) merged confocal image of panels (A, B and D) a confocal image merged from field image the (bright C). (E) the Illustration of emission spectra for TPE-Cy at different pH and buffer in the presence of 1,2-dioleoyl-glycero-3-phosphocholine (DOPC). (F) A plot emission intensities vs pH of solution at 489 and 615 nm respectively. *Reprinted from Chen S, Hong Y, Liu Y, Liu J, Leung CWT, Li M, Kwok RTK, Zhao E, Lam JWY, Yu Y, Tang BZ. J Am Chem Soc. 2013;135:4926–4929 with permission of The American Chemical Society.*

in pH detection with reversible nature in acid/base equilibrium (Fig. 14). According to the proposed mechanism the luminescent properties of the compound was effectively influenced by the protonation and deprotonation process and therefore it is considered as an important component for pH sensing. Further the compound is investigated at different pH in an aqueous solution. The synthesized benzothiazole compound was used for fluorescence imaging and for studying the intracellular pH sensing using Hela cells (Fig. 15).

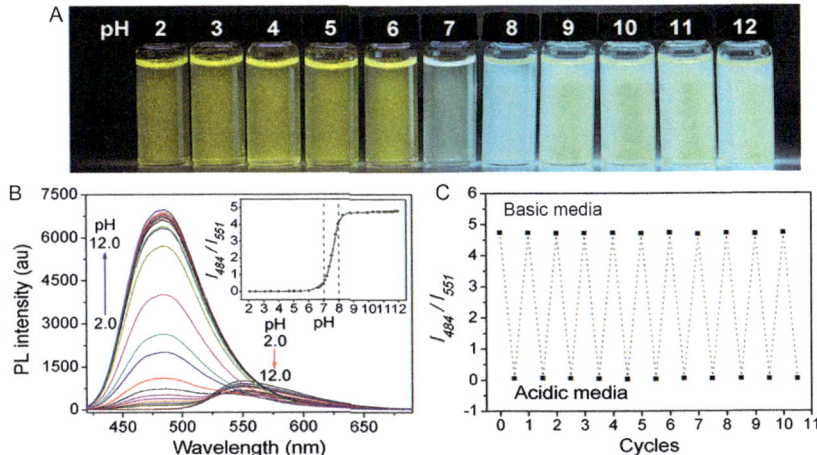

Fig. 14 (A) Fluorescence Image. (B) Illustration of fluorescence emission plot of **3** compound at different pH in water/ethanol fraction. Inset: represents the fluorescence intensity ratio vs pH concentration 50 μmol/L (C) A plot demonstrating reversibility in fluorescence changes for compound in acidic and basic environment. *Reprinted from Li K, Feng Q, Niu G, Zhang W, Li Y, Kang M, Xu K, He J, Hou H, Tang BZ. ACS Sensors. 2018;3:920–928 with permission of The American Chemical Society.*

Lin and coworkers[31] designed a novel dual emission pH-sensitive AIE active fluorescent probe 4,4′-(hydrazine-1,2-diylidene-bis(methanylylidene))-bis(3-hydroxybenzoic acid) (HDBB). This fluorescent probe shows different fluorescence emission at different pH. It showed noticeable fluorescence emission color change from green fluorescence in alkaline medium via subsequent deprotonation, however, in acidic condition displays the orange emission in the aggregated state. So far a number of different AIE-based chemosensor and biosensors have been designed and synthesized for various purpose generally TPE-Cy AIE active molecule frequently used for sensing pH which is designed with AIE active TPE unit and hemicyanine dye which is considered as an highly pH-responsive moiety. New strategy based on Schiff base was designed to construct AIE-based pH fluorescent sensor e.g.: salicylaldehyde Schiff's base azines as pH sensor for cell imaging. The fluorescent molecules show good AIE property. However, the synthesized HDBB derivative showed prominent response to pH in wide range from 2 to 13. More interestingly, it was observed that at low pH HDBB dye showed intense orange luminescence at 590 nm. However, at pH less than 4 fluorescence intensity nearly remains constant and above pH 4 orange emission gradually fades and at pH 7 there is disappears observed (Fig. 16).

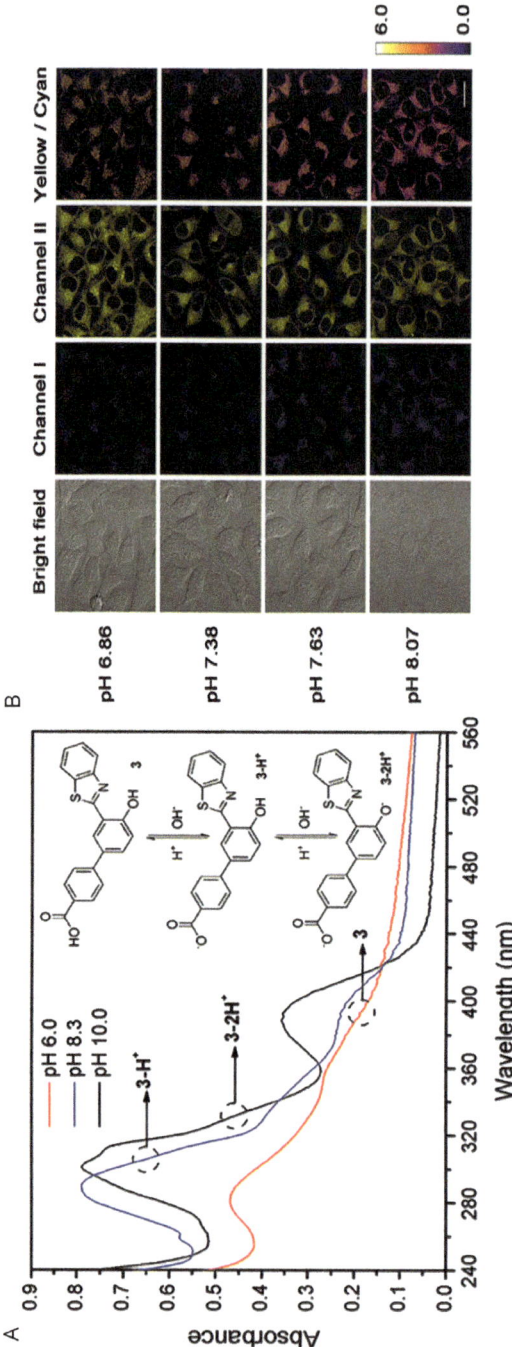

Fig. 15 (A) A spectra illustrating the absorption spectra of the compound at three different pH. Inset represents the protonation and deprotonation sensing mechanism for compound 3 at 50 μmol/L and (B) Illustration of confocal images at different pH buffers on incubation of HeLa cells at 50 μmol/L compound. Reprinted from Li K, Feng Q, Niu G, Zhang W, Li Y, Kang M, Xu K, He J, Hou H, Tang BZ. ACS Sensors. 2018;3:920–928 with permission of The American Chemical Society.

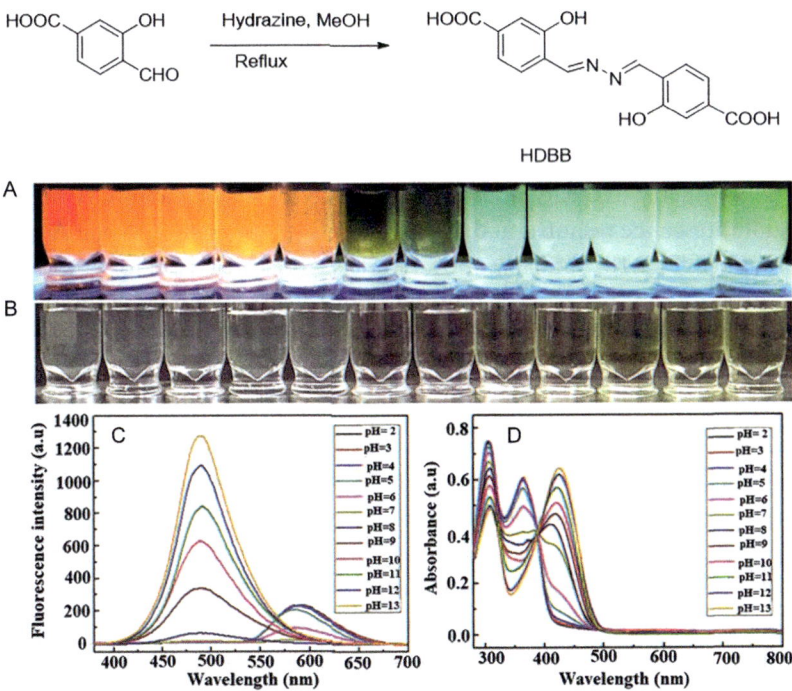

Fig. 16 Synthetic scheme of compound HDBB. (A) The fluorescence image of probe HDBB under UV-light irradiation in aqueous solution at different pH (2–13) at 365 nm. (B) Images under daylight. (C) A plot illustrating the pH dependent fluorescent spectra at different pH from (2–13) for HDBB probe measured at 365 nm in aqueous solution. (D) Plot illustrating pH dependent absorbance change from pH (2–13) for HDBB probe at 365 nm in aqueous solution. *Reprinted from Lin N, Chen X, Yan S, Wang H, Lu Z, Xia X, Liang M, Wu YL, Zheng L, Cao Q, Ding Z. RSC Adv. 2016;6:25416–25419 with permission of The Royal Society of Chemistry.*

The novel fluorescent probe was found to show applicability in sensing pH in-vitro and in-vivo in HepG2 human liver cancer cell and nude mice upon incubation of HDBB with HepG2 cells for 10 min. The molecule has good cell permeability thus could show strong localized fluorescence with Lysotracker representing the specific intracellular location of lysosomes. Because of low pH 6.0 lysosomes can be easily traced from other cell organelle showing the enhanced fluorescence signal due to AIE character. The pH value was around 5.09 using the HDBB as the pH indicator. This novel HDBB can be used for clinical diagnosis of cancer and in pathology laboratories as low pH of lysosomes in tumor than that of the normal tissues in the cellular environment.

Wang[32] and group developed three ratiometric near-infrared probe for precise determination of intracellular pH in living cells. A fluorescent probe A, B and C were synthesized consisting of tetraphenylethylene (TPE) as donor moiety and hemicyanin as an acceptor group via pallidum catalyzed Suzuki coupling reaction (Fig. 17). This A, B and C probe shows good AIE active property in ethanol/water from 0% to 99% v/v fraction. Absorption and fluorescence response at different pH was examined for probe **A** as represented in Fig. 18. Similarly optical studies were carried out for probe **B** and **C**. Probe A, B, and C, were further applied to investigate the membrane permeability at various concentration of probe and to explore its unique property for determining the intracellular pH by cellular imaging in Hela cells.

All AIE probes showed outstanding cell permeability and accumulate in a same cellular compartment in live cells. At pH 3.0–7.0 fluorescence cell imaging of HeLa cells were carried upon incubation with 15 μM fluorescent probe A, B and C in presence of 5 μg/mL nigericin.

Furthermore, Ma and coworkers[33] prepared a ratiometric fluorescent salicylaldehyde azines (SAs) pH probe containing (NO_2, F, and Cl) as electron accepting substituent and (OMe and NEt_2) as electron donating substituent. The probe comprise of AIE active effect with extended π conjugated naphthalene system and due to chemical reactivity of phenol toward OH^-/H most of the SAs are used in wide range of pH from (2–14) as shows in Fig. 19. Furthermore the synthesized SAs have been used for sensing of various analytes such as Zn^{2+}, Co^{2+}, Cu^{2+}, Fe^{3+}, F^-, CN^- proteins and heparin. Among reported SAs excluding 3-OMe and 4-NEt containing molecules all other derivatives exhibit AIE effect in the organic solvent and water.

Phenol being weakly acidic with reported pKa of 10 thus its salicylaldehyde derivatives with turn ON-OFF fluorescence strategy are used for pH sensing. The pH dependent absorption and fluorescence studies were carried out for the SAs derivatives in mixed MeCN/aq. Britton Robinson (B.R) buffer solution. It was observed at constant pH <7.0 the red emission intensity appears at 565 nm because of protonated 3-Cl further as the pH increases from 7.0 to 11 the emission intensity decreases reducing the emission intensity and lastly remains constant at pH >11 blue-green I_{515} belonging to the protonated state (Fig. 20). Similar pH studies were carried out for other derivatives. Further pH study was also done by using pH test paper.

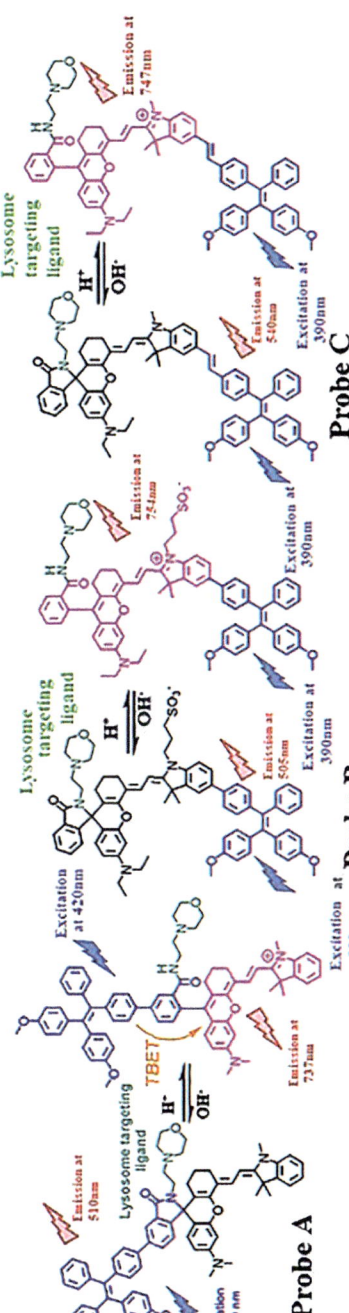

Fig. 17 Chemical structures of the three infrared probe A, B and C illustrating the pH change. *Reprinted from Wang J, Xia S, Bi J, Fang M, Mazi W, Zhang Y, Conner N, Luo FT, Lu HP, Liu H. Bioconjug Chem. 2018;29:1406–1418 with permission of The American Chemical Society.*

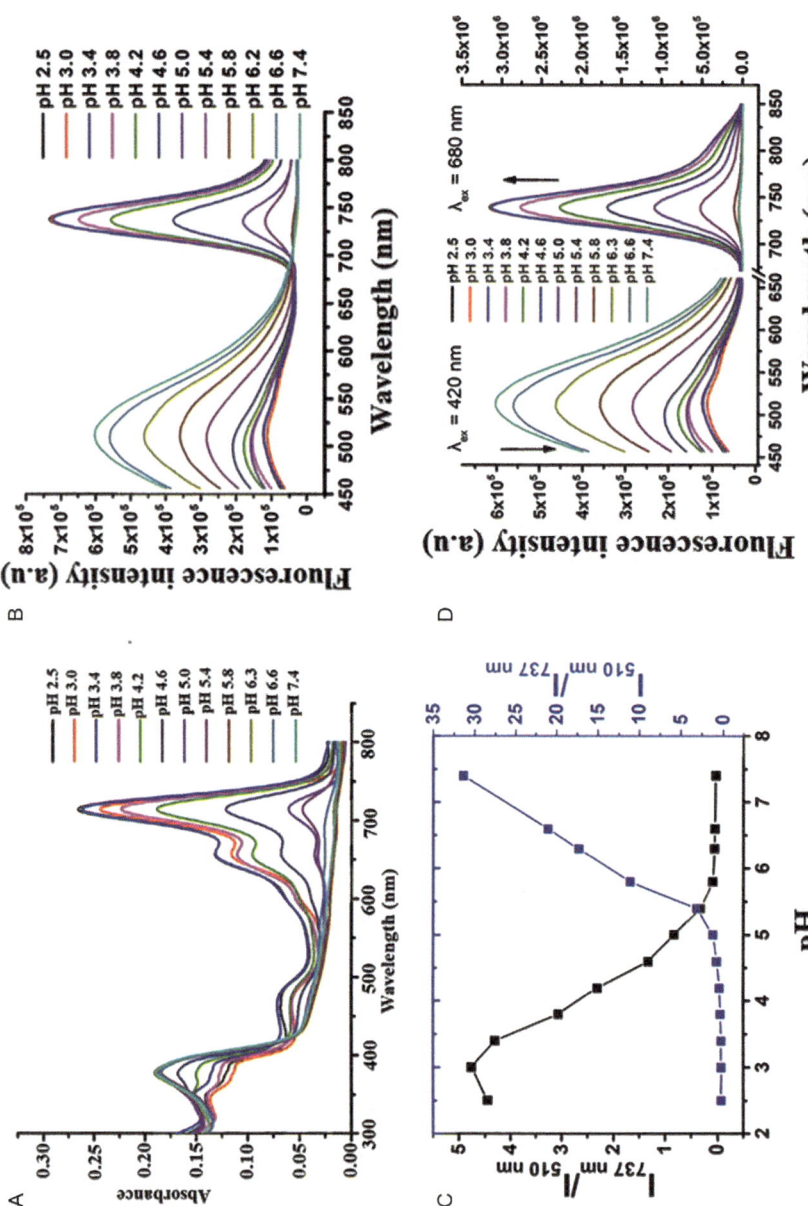

Fig. 18 (A) Graphical representation of absorption spectra for Probe A (5 μM). (B) Illustration of the fluorescence spectra at different pH in citrate buffer for probe A (5 μM) 30% ethanol at λ_{ex} = 420 nm (C) a graph illustrating fluorescence ratios v/s pH of hemicyanine acceptor (I_{737} nm) to TPE donor (I_{510} nm) under λ_{ex} = 420 nm. (D) The graph illustrating the fluorescence intensity at different pH for Probe A at wavelength of 390 and 680 nm. Reprinted from Wang J, Xia S, Bi J, Fang M, Mazi W, Zhang Y, Conner N, Luo FT, Lu HP, Liu H. Bioconjug Chem. 2018;29:1406–1418 with permission of The American Chemical Society.

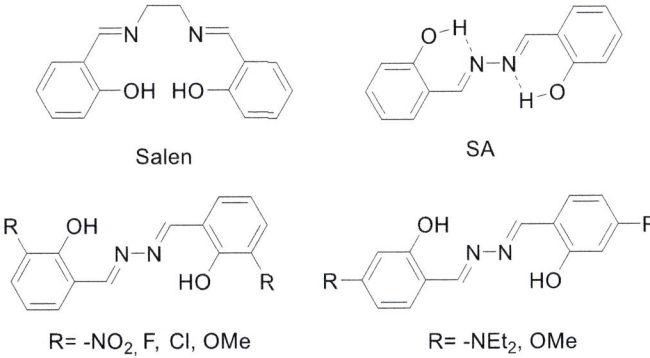

Fig. 19 Structural representation for Salen and SAs fluorescent derivative.

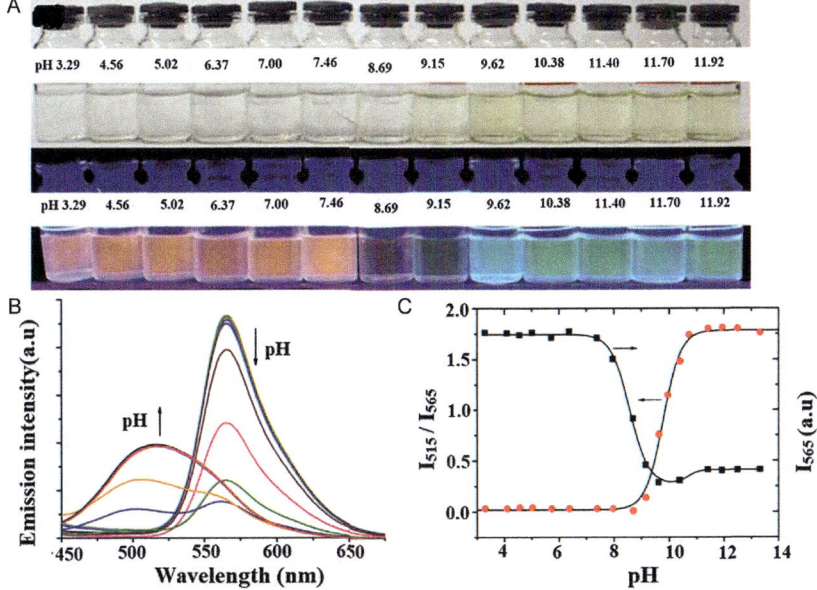

Fig. 20 Images describing color change (A) top under Visible light and bottom under UV light illumination at 365 nm in different pH range from 3 to 12. (B) Represents the emission spectra upon variation in pH. (C) Demonstration of relative plot of emission intensity vs pH value for 3-Cl in B.R buffer. *Reprinted from Ma X, Cheng J, Liu J, Zhou X, Xiang H. New J Chem. 2015;39:492–500 with permission of The Royal Society of Chemistry.*

Our group reported a an AIE active tetraphenylethylene based (Py-TPE) functionalized with pyridine moiety. In this present work, pyridyl functionalized dye is used as pH sensor and acid-induced assembly was well studied.[34] Py-TPE is used as the reversible probe for acid/base sensing and to study the pH and solvent dependent self-assembly in solution and live cells. Py-TPE was synthesized by the reacting tetra-bromo TPE with 4-pyridine boronic acid in presence of $Pd(PPh_3)_4$ catalyst via Suzuki coupling. Absorption, emission, and naked-eye detection can be easily done by using Py-TPE in different organic solvents (MeOH, DMF, $CHCl_3$). A naked eye detection is possible for the probe due to color changes occurring in the solution by protonation/deprotonation process. The protonation and deprotonation process is investigated by addition of acid and base and color change was observed for Py-TPE probe from light yellow to dark green in acidic condition by the addition of tetrafluoro acetic acid (TFA) and upon immediate addition of triethylamine base dark green color reverses back to original light yellow in DMF, methanol and chloroform as represented in Fig. 21.

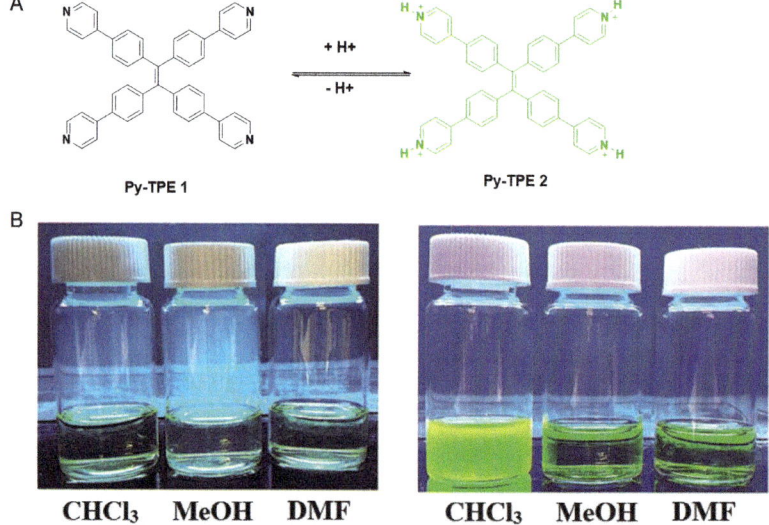

Fig. 21 (A) Schematic presentation of protonation/deprotonation mechanism for Py-TPE molecule (1 and 2) (B) Photographs of Py-TPE (1 and 2) under UV light illumination ($\lambda_{ex} = 365$ nm) at (0.5 µM) in presence and absence of TFA (2 µM) at room temperature in different solvent system. *Reprinted from Rananaware A, Bhosale RS, Patil H, Al Kobaisi M, Abraham A, Shukla R, Bhosale S V., Bhosale S V. RSC Adv. 2014;4:59078–59082 with permission of The Royal Society of Chemistry.*

During the protonation and deprotonation cycle, it was observed that Py-TPE shows weak fluorescence ($\lambda_{em} = 480$ nm, $\lambda_{ex} = 340$nm) in various solvent system such CHCl$_3$, DMF, and MeOH. Py-TPE markedly shows high fluorescence solid state (i.e., band at 438 and 562) and in water (at 425 and 550 nm). Furthermore, it was observed that Py-TPE upon protonation emission band broadens with redshift to 440 and 540 nm. In chloroform, it was observed that emission was maximum at 440 nm but red-shifted to 445 nm in MeOH and DMF. Enhancement of fluorescence intensity takes place upon the addition of TFA in CHCl3, DMF, MeOH. The process is reversible and fluorescence can be restored back to its original state after the addition of triethylamine (TEA). The intramolecular charge transfer and π-π stacking of the aromatic core are responsible for clear bathochromic shift and enhanced emission for protonated Py-TPE over the deprotonated species. Needle shaped crystal of protonated Py-TPE were formed by the self-assembled aggregate arrangement. This fluorescent Py-TPE 1 is used intracellular pH sensing application. The solution of Py-TPE 1 was prepared in DMSO at a 0.5 mg mL^{-1} human prostate cancer (PC-3) were used for the bioimaging. These cells were pre-treated for 2 h with the 0.5 μg mL^{-1} of Py-TPE 1 and later washed extensively in order to remove excess amount of unreacted Py-TPE 1 with ice-cold PBS. Further at low temperature the fluorescence of the cell was investigated in acidic neutral and alkaline condition (pH 3,7 and 9 respectively). However, enhanced fluorescence observed for cells in blue channel at 450 and 490 and 470 nm a neutral pH condition but fluorescence remains un-altered with increase in to pH 9 (Fig. 22).

Wang group reported a simple fluorescent probe comprising of coumarin as electron donor and quinoline electron acceptor by linking N,N' deformyldihydrazine to give 2-oxo-N-(2-quiolin-8-yloxy)acetyl-2H-chromene-3-carbohydrazide (CHBQ) (Fig. 23A).[35] The CHBQ probe shows good AIE active property in acetonitrile and water due to which it is considered a facile probe for detecting pH in the microenvironment. It was observed that the absorption changes of the probe were found to show the shoulder peak at 330 nm gradually decreases as the pH increases and gets red-shifted to 370 nm. The fluorescence studies showed that at fluorescence of CHBQ decreases in the pH ranging from 2.00 to 6.00 at 475 nm with appearance of red-shift at 415 nm. Further increase in the pH from 7 to 12 there is a gradual decrease in fluorescence of CHBQ. In order to study the sensitivity of CHBQ probe for determining pH acid and alkali titrations are performed. The nonlinear fit of the sigmoidal curve in acid and alkali

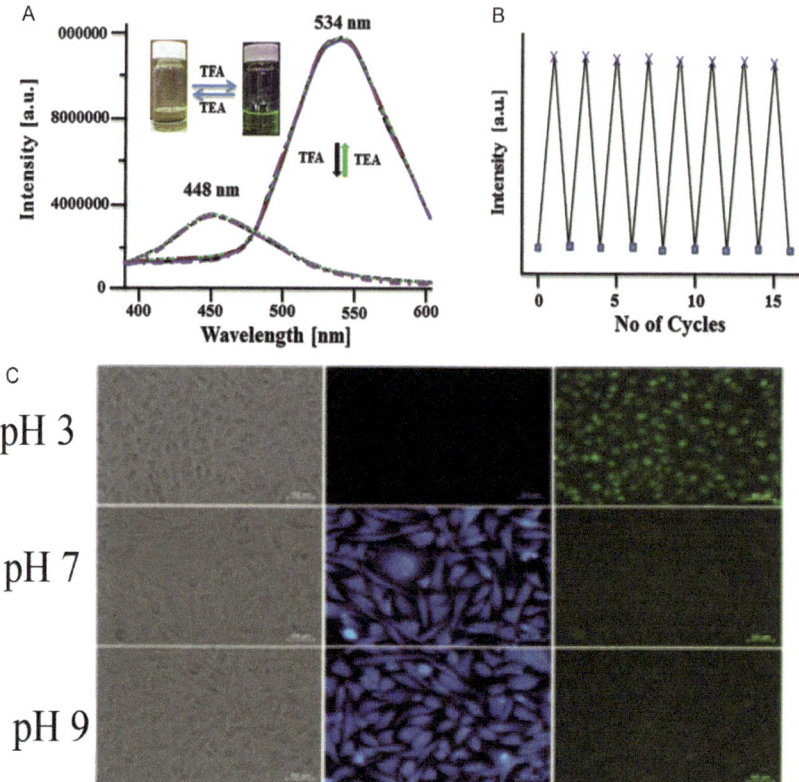

Fig. 22 (A) PL spectra of addition of TFA and TEA in **1** (1×10^{-5} M) in DMF ($\lambda_{ex} = 365$ nm). (B) A plot at 534 nm illustrating fluorescence intensity vs number repeated cycles upon additions of TFA and TEA. (C) Confocal fluorescence microscopic images of Py-TPE probe with (5 µg mL^{-1}) for PC-3 cells treated for 2 h at different pH condition. *Reprinted from Rananaware A, Bhosale RS, Patil H, Al Kobaisi M, Abraham A, Shukla R, Bhosale S V., Bhosale S V. RSC Adv. 2014;4:59078–59082 with permission of The Royal Society of Chemistry.*

which afford pKa value of 4.26 and 8.60. pH detection of probe CHBQ was also carried out by using paper strips which is a simple device for broad range pH detection. The color of test strip changes to cyan visualized under UV-light illumination at 365 nm when the drop of acidic solution is placed on the paper strip of CHBQ probe. Gradual increase in the pH changes the color of the test strips to blue. However, further increase in pH decrease the blue color of the test strips which can be observed in the naked eye under UV lamp at 365 nm. This paper strips of CHBQ probe detects the volatile solvents such as acetic acid and ethylenediamine (Fig. 23B).

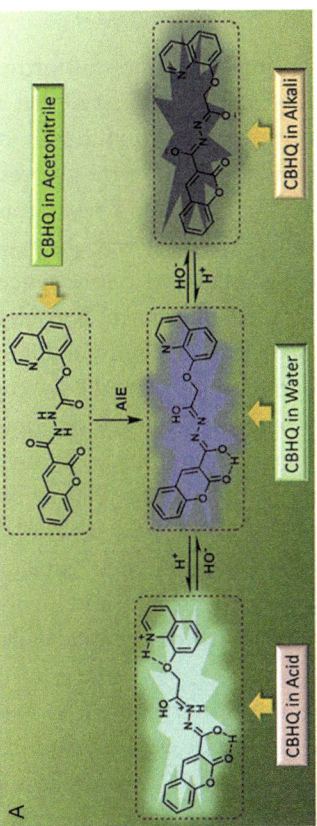

Fig. 23 (A) Representation of structure of the compound CHBQ in acidic, water and in alkali medium (B) image of test strips CHBQ probe representing the color change under UV$_{365\,nm}$ lamp at different pH value. *Reprinted from Wang X, Wang H, Niu Y, Wang Y, Feng L. Spectrochim Acta A Mol Biomol Spectrosc. 2020;226:117650 with permission of Elsevier.*

Qui and coworkers designed and synthesized AIE active and FRET based conjugated cyano-diphenylethylene BODIPY pH sensor.[36] The compound was synthesized by condensation of 4-hydroxybenzene acetonitrile and 4-(diethoxymethyl)benzaldehyde. Due to FRET and AIE effect compound exhibited intense fluorescence in THF/H$_2$O mixture. The compound shows good fluorescence response to pH change with pKa value of 9.79 and can be used for the detection of CO_3^{2-}. The compound shows good AIE active property and was successfully used for pH study detection (Fig. 24).

Since the compound can be used for pH sensing the probe was further studied for the fluorescence reversibility at pH 7–11. It was observed that at pH 7 the solution of the probe was highly fluorescent but when the pH of the same solution was increase to pH 11 by adding of NaOH the fluorescence completely quenches. After that to the same solution a few drops of conc. H$_2$SO$_4$ was added to bring the pH back to 7 and it was observed that the solution of the probe showed again the enhanced fluorescence. These phenomena of Turn-on-Turn off was carried out in 5 cycles. These show that the probe has excellent fluorescence reversibility in different pH range. As the compound explored its applicability for pH sensing hence the probe acts as good fluorescent marker for in vitro cell imaging. It can be seen that the MCF-7 cells upon incubation with probe at different pH 7.0, 8.0, and 9.0 showed bright fluorescence. However, the cells without and with incubation in the compound at pH 7.0–11.0 show very weak fluorescence or no fluorescence response. This phenomenon clearly explains that living cells with compound show fluorescence change and hence BODIPY sensor can be effectively utilized for determination of intracellular pH in live cells.

Wang and coworkers developed another novel tetraphenylethylene (TPE) monomer base triple mode fluorescent pH probe which sense pH of solution selectively in wide pH range (1.99–11.69).[37] The molecule shows different photophysical changes at different pH condition, i.e., emission in acidic condition, while exhibit ratiometric fluorescence under the neutral condition and finally in basic condition exhibit aggregation-induced emission. The target molecule was synthesized as shown in (Fig. 25). The probe showed good AIE active property hence can be effectively used for pH sensing. When the probe was dissolved in the acidic condition the amine moiety gets transformed into ammonium salts at pH (1.99–7.20) fluorescence quenching of monomer takes place at 377 nm. Moreover, a new fluorescence band appears in alkaline condition as the pH increases (7.02–8.50)

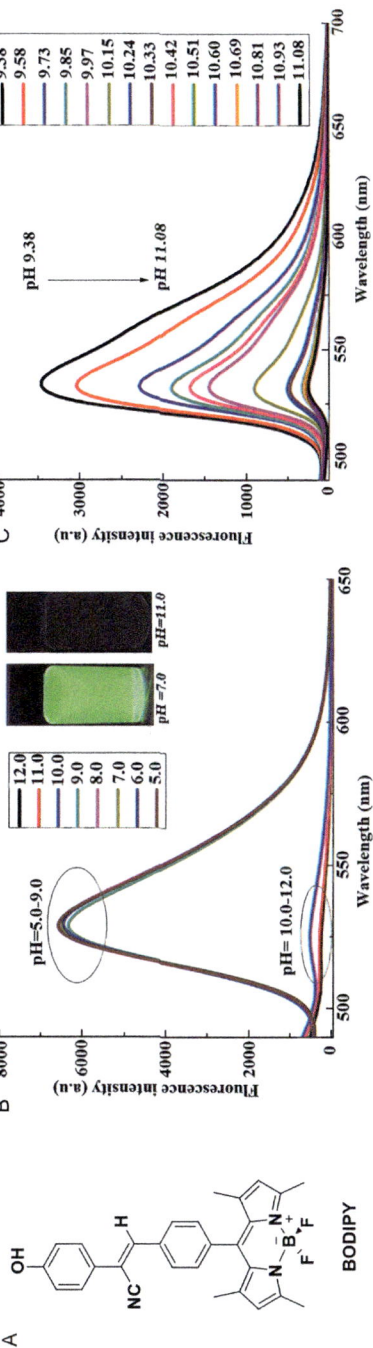

Fig. 24 (A) Chemical structure of the BODIPY probe. A plot of fluorescence emission spectra of the BODIPY in THF/H$_2$O mixtures at 370 nm, Inset: photographs in under UV. (B) Left: at pH = 5.00–12.00. (C) Right: at pH 9.38–11.08 in buffer solution. *Reprinted from Qiu J, Jiang S, Guo H, Yang F. Dye Pigment. 2018;157:351–358 with permission of Elsevier.*

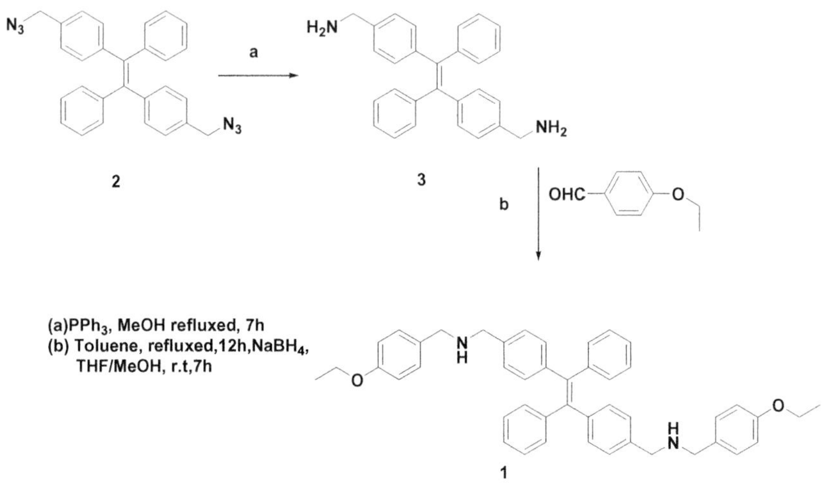

Fig. 25 Illustration of the synthetic route of the probe **1**.

due to AIE effect at 483 nm. Further increase in the pH from (8.50–11.04) AIEE band was observed at 483 nm. Hence this probe showing the AIE process provides good utilization for pH detection.

Hence a constructed new amphiphilic fluorescent probe **1** for sensing pH containing TPE as a hydrophobic moiety linked with secondary amines as hydrophilic moieties. The probe **1** exhibit good AIE active property where the probe initially has weak fluorescence. However, the fluorescence intensity increases as the water fraction increase from 0% to 80% water in THF. The absorption titration was carried out in THF/water at (2:8 v/v) for the probe **1** where the pH value changes from 2 to 8 there is little change in two wavelengths at 277 and 328 nm. Further, no observed in strong alkaline condition (9.94–11.84) at 328 nm with fluorescence enhancement at 277 nm. pH titration of the probe in THF/H_2O (2:8 v/v) as shown in (Fig. 26) was carried out at different pH condition as the pH increases the fluorescence intensity decreases at 377 nm. It was observed that in alkaline solution there is a gradual decrease in emission intensity at 377 nm along with the appearance of a new emission band at 483 nm. The fluorescence intensity increases as the pH increases 8.50–11.64 at 483 nm. Intense fluorescence observed in alkaline condition while the fluorescence completely disappears at low pH. Probe **1** is pH dependent thus acts as a typical molecular ON-OFF switch. TEM image revealed that in acidic pH the aggregate formed has the spherical shape with approximate diameter of 2 µM, however, in basic condition TEM images revealed that probe **1** shows hydrogen bond J-aggregation and self-assembles to give linearly crystalline TPE nanowire of 200–450 nm diameter.

Particularly in the recent years, a multi responsive fluorescent polymer-based nanoparticle has gain more importance toward temperature and pH sensing. Most of the polymeric hydrogels exhibit the highly sensitive response to temperature and pH.[38] Herein, Zhao[39] and coworkers designed a hydrogel nanoparticle by incorporating rare earth Eu (III)(TTA)$_3$ Phen complex in to PNIPAM polymer and quaternary ammonium tertraphenylethylene derivative (d-TPE) doped PNIPAM-co-PAA. The synthesized hydrogel nanoparticle found to be strongly and very promising independent candidate as it shows dual emissive response to pH and temperature (Fig. 27). The hydrogel containing the hydrophobic core shows red emission which is pH-independent hence can be utilized fluorescent marker to trace the cells while hydrophilic shell exhibit blue emission which is highly sensitive to pH. Hence the cancer cell can be easily detected by the synthesized hydrogel nanoparticle. The PL intensities exhibited linear response to temperature in range from 10 °C to 80 °C. However, blue emission for shell showed the linear response in neutral pH range from 6.5 to 7.6.[40]

The two PL emission peaks observed at 613 and 468 nm for Eu(TTA)$_3$Phen and d-TPE hydrogel molecule respectively. Moreover the pH responsive study was performed for hydrogel nanoparticle where it can be used to distinguish the normal cell (7.4 pH) with cancer cells (6.5 pH). It is revealed that at pH (4.0–7.6) the electrostatic interaction becomes stronger between PNIPAM-co-PAA and positively charged d-TPE molecule as the degree of ionization increases of Eu-doped PS-co-PNIPAM/PNIPAM-co-PAA nanoparticle. So, as the pH increases the PL emission intensity at 468 nm also increases. But at low pH, the electrostatic interaction becomes weaker between PNIPAM-co-PAA and positively charged d-TPE molecule which leads to dissociations of d-TPE freely from PNIPAM-co-PAA carboxylic group hence decreases in PL intensity with decreasing pH (Fig. 28). The hydrogel demonstrated the thermo-responsive characteristics which showed the temperature variation PL intensities. The two emission bands appears at 468 and 613 nm for d-TPE molecule and Eu(TTA)$_3$Phen respectively. The photoluminescence intensity decrease linearly with increasing temperature from 10 °C to 80 °C, i.e., the PNIPAM-co-PAA shell/core restricts the intramolecular rotation of d-TPE and shrinks with the increasing temperature (Fig. 28). Due to these PL intensity variations with pH and temperature make this nanoparticle to be used for cell imaging and tissues.

Feng[41] and group synthesized Schiff base 4-N,N-dimethylaminoaniline salicylaldehyde (DAS) AIE active derivative. In solid aggregative state the

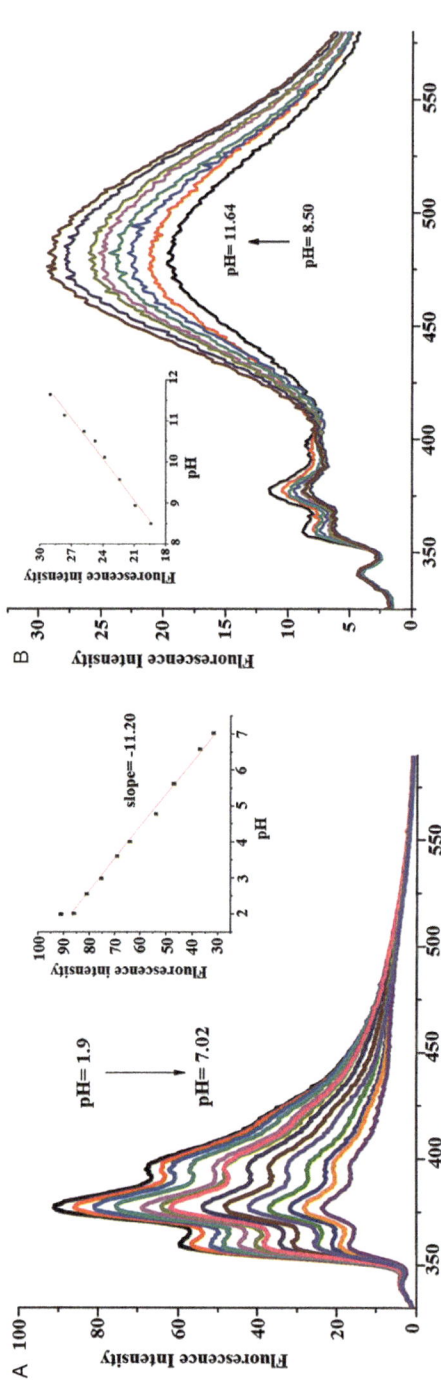

Fig. 26 (A) Fluorescence spectra of probe 1 (10 μM) at different pH from 1.99 to 7.02 in THF–H_2O fraction. Inset: Plot of fluorescence intensity vs pH (1.99–7.02) at 377 nm. (B) Fluorescence spectra in pH range from 8.25 to 11.64 in THF-Water Inset: representing plot of fluorescence intensity vs pH at 483 nm. *Reprinted from Wang Z, Ye JH, Li J, Bai Y, Zhang W, He W. RSC Adv. 2015;5:8912–8917 with permission of The Royal Society of Chemistry.*

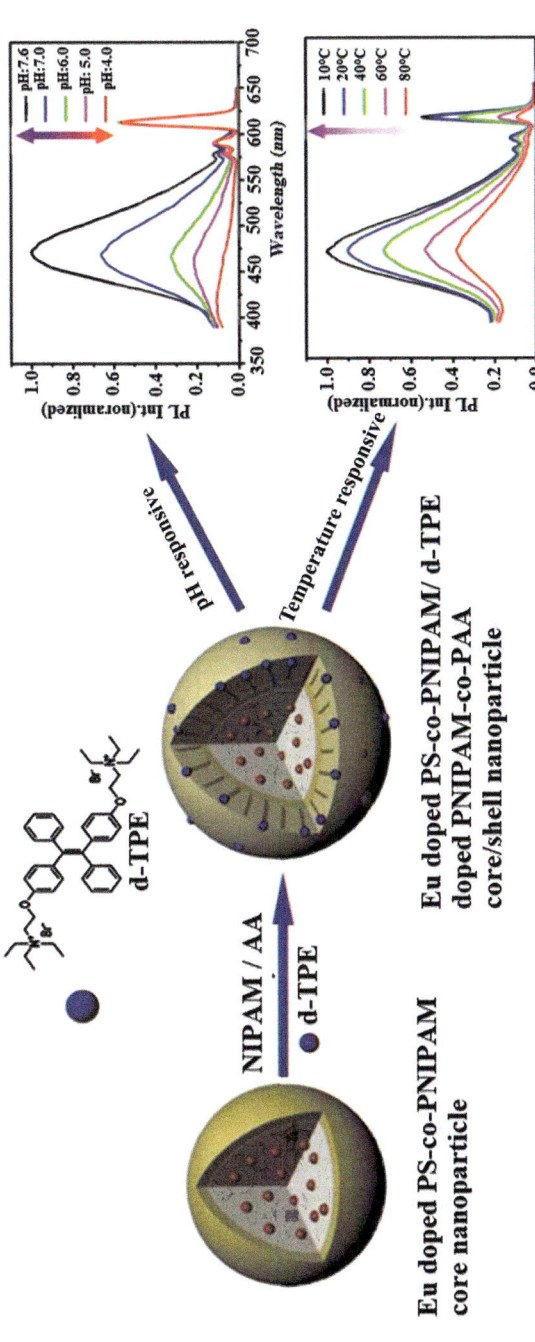

Fig. 27 The schematic representation of synthetic route of Eu-doped PS-co-PNIPAM/d-TPE doped PNIPAM-co-PAA core/shell hydrogel nanoparticles representing the dual mode emission response to the temperature and pH. *Reprinted from Zhao Y, Shi C, Yang X, Shen B, Sun Y, Chen Y, Xu X, Sun H, Yu K, Yang B, Lin Q. ACS Nano. 2016;10:5856–5863 with permission of The American chemical Society.*

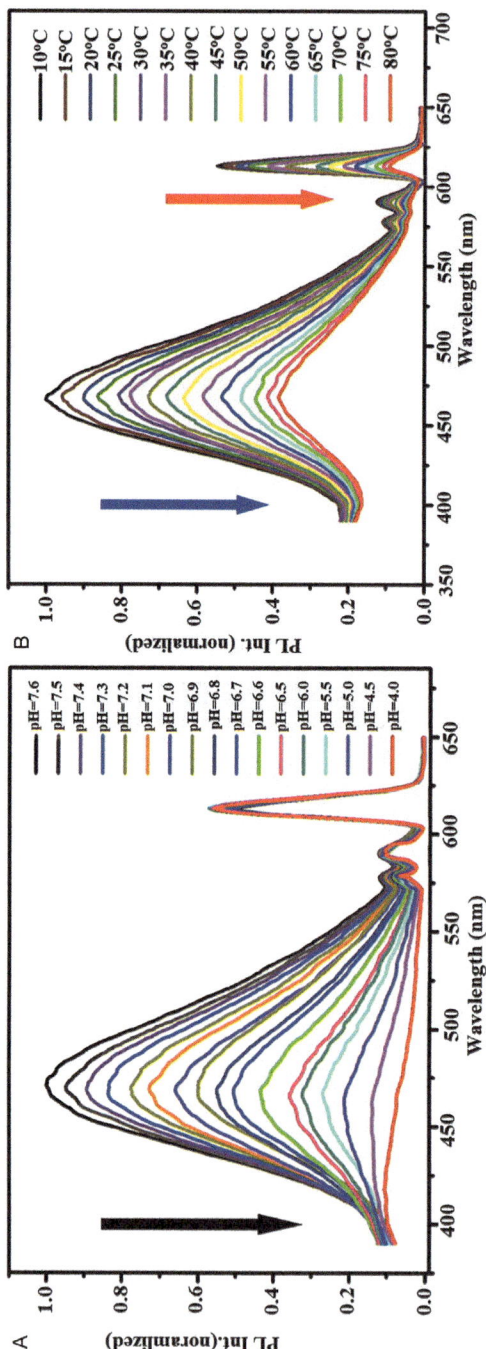

Fig. 28 (A) illustration of PL plot for Eu-doped PS-co-PNIPAM/d-TPE-doped PNIPAM-co-PAA hydrogel nanoparticle at different pH values (λ_{ex} = 360 nm). (B) Illustration of PL spectra at different temperatures (λ_{ex} = 360 nm). *Reprinted from Zhao Y, Shi C, Yang X, Shen B, Sun Y, Chen Y, Xu X, Sun H, Yu K, Yang B, Lin Q. ACS Nano. 2016;10:5856–5863 with permission of The American chemical Society.*

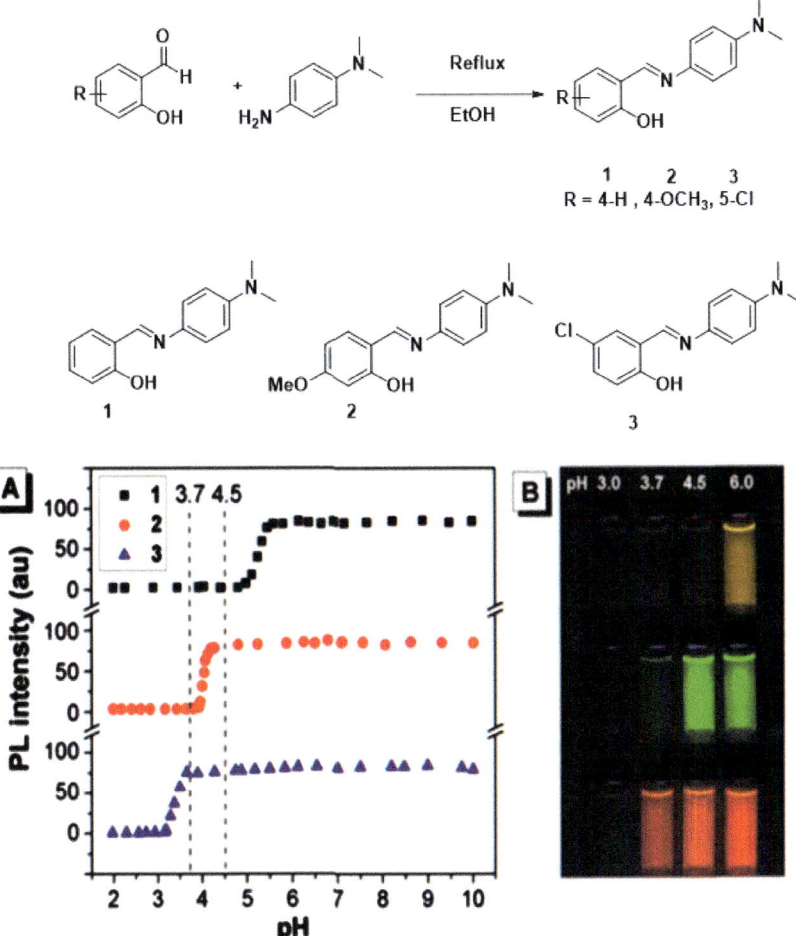

Fig. 29 Schematic illustration of synthetic route for the synthesis DAS (A) PL spectra of fluorescence emission vs pH for **1–3**. (B) Image demonstrating fluorescence change at different pH value from 3.0 to 6.0 of **1–3** with excitation and emission wavelengths (333/542, 338/513, 356/580 nm) respectively under UV light illumination. *Reprinted from Feng Q, Li Y, Wang L, Li C, Wang J, Liu Y, Li K, Hou H. Chem Commun. 2016;52:3123–3126 with permission of The Royal Society of Chemistry.*

derivative exhibit good fluorescence emission property with high fluorescence quantum yield. Due to its excellent pH-dependent optical properties this Schiff's bases (DAS) explored its potential application in pH sensing. The pH-dependent fluorescence studies for 1–3 in aqueous solution was carried out at different pH from 2.0 to 10.0. The compound demonstrated different fluorescence in alkaline and acidic medium, i.e., exhibiting highly intense fluorescence emission in alkaline condition and weak fluorescence in acidic condition (Fig. 29).

4. AIE active molecules for viscosity sensing

As the temperature and pH are considered to be the most important factors in intracellular processes which can be monitored by using fluorescent materials similarly viscosity is also an important parameter for determining the viscosity of biofluids from the cellular level (e.g., in cell signaling changes in membrane and cytoplasmicviscosity) to organism level (e.g., to study the viscosity change in blood plasma or lymphatic fluid, in diabetes, hypertension, infraction, and aging).[42,43] Viscosity measurement of biofluids provides the excellent tool for diagnosis of various infectious diseases and pathologies.[44–46] The conventional mechanical methods such as capillary viscometer, falling base viscometer, and rotational viscometer are cumbersome to use and incapable for real-time measurement of viscosity.[47] Therefore, there is need of developing methods which are applicable at the molecular and cellular level. One of the best approaches accomplished by the use of various kinds of fluorescent material, molecular rotors[48,49] AIE active molecules and dyes which are much more effective in investigation of intracellular viscosity.[50]

Quantum yield of viscosity dependent fluorescent emission for fluorescent molecules can be characterized and well described by Forster Hoffmann Equation.[51]

$$\Phi_f = Z \eta^\alpha$$

where Z and α are constants

η = viscosity

Hoffmann Forster equation can be used to find Φ_f quantum yield and τ_f fluorescence life time of molecular rotors.[52] In our book chapter we are focusing on AIE active molecules that can be used for sensing viscosity.

Here in Kumbhar[53] and coworker synthesized the carbazole styryl based fluorescent molecular rotors (FMR)[54] by functionalizing C=C bond with different chemical derivatives. The molecule was studied for its effect on change in optical properties, AIE characteristics and to study its viscosity parameter of molecular rotors that take place upon functionalizing the styryl derivatives. These compounds displayed highly sensitive response for a viscous medium. In low viscous medium the molecule undergoes free rotation along C=C bond but free rotation along C=C are restricted upon increase in viscosity of the medium. The molecules undergoes free rotations in low viscous medium which shows the excited-state relaxation by a non-radiative

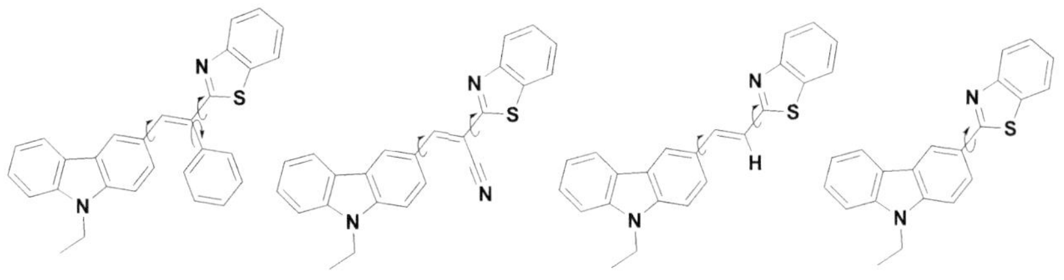

Fig. 30 Illustration of carbazole styryl based FMRs.

pathway. However highly viscous microenvironment restricts intramolecular rotation and excited-state relaxation occurs via enhancement in fluorescence emission. As the size of the molecules increases greater will the viscosity sensitivity which results in the formation of aggregate resulting in restriction of intramolecular rotation thus enhancing the fluorescence. This phenomenon is known as aggregation-induced emission (AIE).[55] Following are the chemical structures of molecular rotors (Fig. 30).

Binol and its derivatives are regarded as an essential source of fluorescent material[56] based on these herein Yang[57] et al. synthesized the axially chiral 1,4 dihydropyridine derivative ((R)-/(S)-2) with aggregation induced emission by combining 1,1-binaphthol (BINOL) and 1,4-dihydropyridine derivatives. (Fig. 31).

(R)-/(S)-2 have weak fluorescence in THF solution but upon aggregate formation the molecule is red shifted with exciplex formation. Thus, shows the linear and multi exponential relationship between fluorescence intensities and viscosity of local excited and exciplex which helps to determine the viscosity of the medium in various mixed solvents. This probe shows excellent application fluorescence imaging and in the field of chiral recognition. The absorption and emission spectra of (R)-/(S)-2 were investigated in different solvents from hexane to glycol. However, it was observed that glycol showed new emission but with stronger fluorescence intensity than that with other solvents. The viscosity dependent emission behavior was studied and revealed that in there occurs intramolecular rotations in viscous solution of (R)-/(S)-2 and exist in two conformations, i.e., in open (O) conformation with extended butylene chain away from BINOL and closed (C) conformation in which the proton present on nitrogen of dihydropyridine and proton on oxygen of BINOL points toward the π-electrons of aromatic system. Thus, providing local excited emission (LE)λ_{em} and exciplex to demonstrates AIE active properties which is good means to quantify the viscosity. Infact this proves that probe can be used

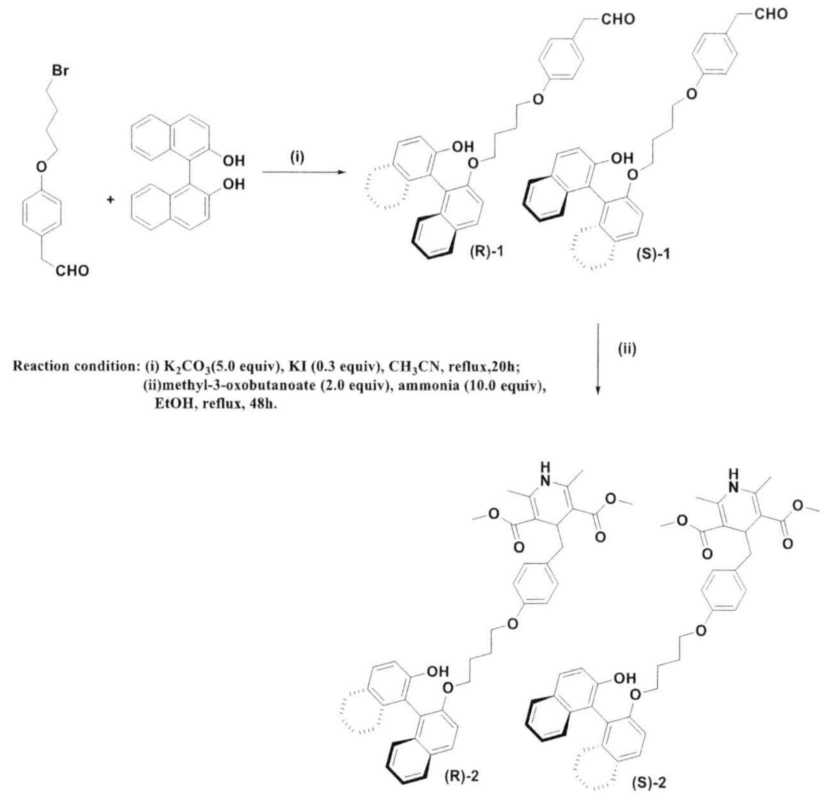

Fig. 31 The schematic illustration for synthesis of **(R)-/(S)-2**.

for viscosity sensing. The viscosity sensitive behavior of (R)-/(S)-2 solution was further studied by adding glycerol. The distinct emission bands were observed for (R)-/(S)-2 with steady increase in the emission intensity for local excited emission λem and exciplex λem as the viscosity is increased by increamental addition of glycerol from 0% to 80%. However, it was experimental noted that the emission intensities of (R)-/(S)-2 shows 2-fold enhancement for local excited emission while 11-fold enhancement in emission intensity for exciplex λem compared to 10% glycerol methanol fraction. So increase in glycerol fraction from 0% to 80% showed the best fitted line with relatively good correlation coefficient. But as the ratio of glycerol in glycerol/methanol mixtures increases from 80% to 99% the emission intensities of (R)-/(S)-2 decreases gradually due to precipitation occurs in viscous solution. Thus making (R)-/(S)-2 to be good fluorescent ratiometric sensor to determine viscosity in glycerol methanol fraction (Fig. 32).

Chen[58] and coworkers demonstrated applicability of TPE-Cy dye for studying the AIE property, cell permeability and mapping the viscosity in

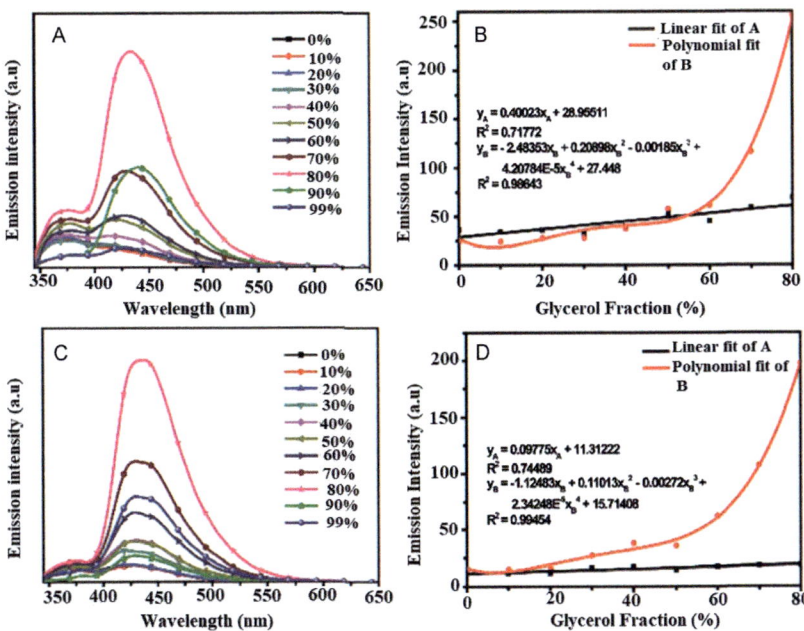

Fig. 32 PL spectra of (R)-2 (A) and (S)-2 (C) in glycerol/methanol (50 µM). (B) Illustration of callibration plot for (R)-2 with emission intensities at 372 and 431 nm (D) illustration of emission intensities for (S)-2 at 375 and 432 nm as the viscosity increases. *Reprinted from Yang Z, Huo Y, Liu Y, Du G, Zhang W, Zhou L, Zhan L, Ren X, Duan W, Gong S. RSC Adv. 2019;9:32219–32225 with permission of The Royal Society of Chemistry.*

live cells. In this molecule, the fluorescence lifetime of the AIE active hemicyanin dye increases with increasing viscosity. The structure of TPE-Cy dye illustrated in the (Fig. 33). The fluorescence behavior concerning viscosity changes was studied by using ethylene glycol and glycerol mixtures in different fraction.(Fig. 33A). It was observed that TPE-Cy-OH shows weak emission in pure ethylene glycol but as the fraction of glycerol increases the viscosity of the solution increases and thus enhances the blue emission of TPE-Cy-OH (Fig. 33B). The molecules shows good AIE phenomenone. Initially the TPE serves to be non-emissive as the phenyl rings freely rotate in solution state. But as the viscosity of the medium increases it inhibit the non-radiative decay and shows the high emissive property. The fluorescence intensity enhanced by 18-fold with 99.8% in glycerol fraction. The fluorescence lifetime is yet another important parameter for viscosity measurement in TPE-Cy-OH so the fluorescence lifetime increases as the viscosity of the solvent increase. The fluorescence lifetime of TPE-Cy was further investigated by using different membrane fluids with different composition of phospholipids and different structures. Herein, four different artificial lipid

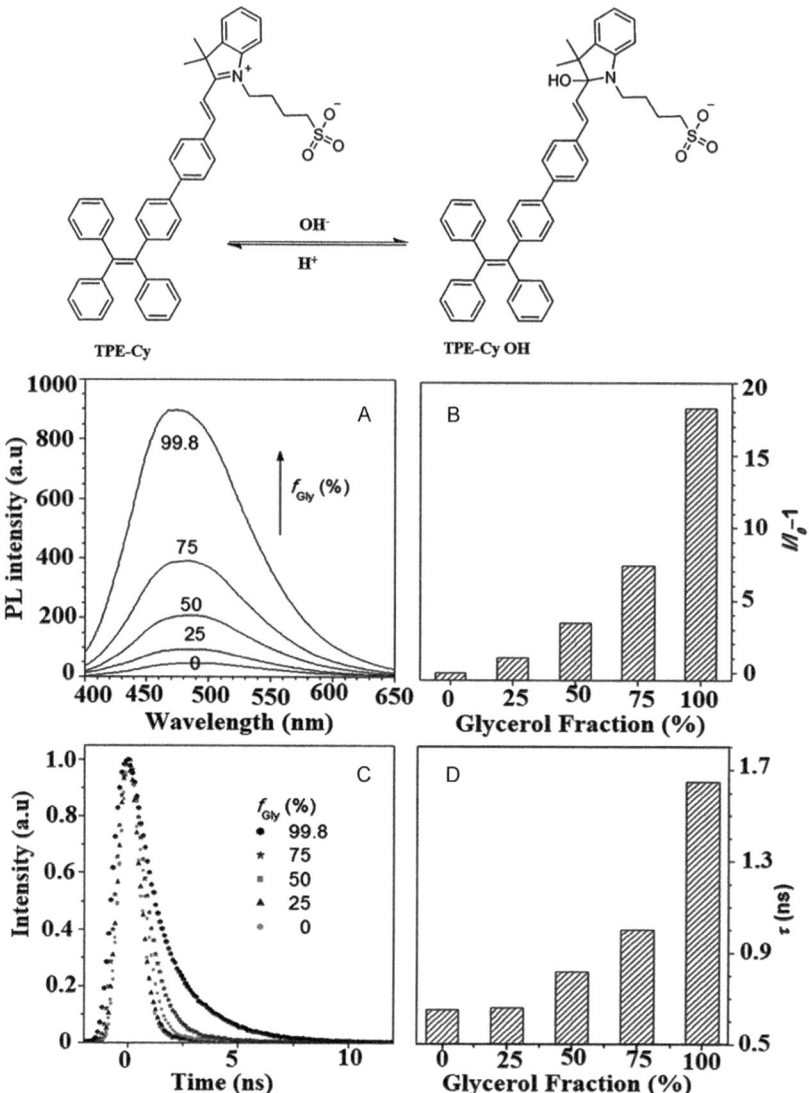

Fig. 33 A schematic representation for TPE-Cy structural transformation to TPE-Cy-OH. (A) Illustration of PL spectra of TPE-Cy-OH in ethylene glycol and glycerol with different glycerol fraction (f_{Gly}). (B) Plot describing fluorescence intensity vs glycerol fraction. (C) In mixture of ethylene glycol and glycerol fluorescence decay trace of TPE-Cy-OH. (D) Plot demonstrating the fluorescence lifetime vs glycerol fraction at $f_{gly}\lambda_{ex} = 375$ nm, and $\lambda_{em} = 480$ nm. *Reprinted from Chen S, Hong Y, Zeng Y, Sun Q, Liu Y, Zhao E, Bai G, Qu J, Hao J, Tang BZ. Chem Eur J. 2015;21:4315–4320 with permission of Wiley-VCH.*

composition and structures were used such as 1,2-dioleoyl-sn-glycero-3-phosphocholine (DOPC), 1,2-diheptanoyl-sn-glycero-3-phosphocholine (DHPC),1,2-distearoyl-sn-glycero-3-phophocholine (DSPC) and cholesterol. It was observed that the shortest lifetime of 0.71 ns when unsaturated DOPC lipid composition was used for TPE-Cy but highest lifetime of 1.36 ns appears in the presence of saturated form DSPC lipid. Further, addition of cholesterol to the DOPC lipid vesicles the lifetime becomes longer than that of pure DOPC lipid. The TPE-Cy exhibit a lifetime between DSPC and DOPC/cholesterol vesicles with DHPC lipid containing small saturated aliphatic lipid chain. Based on viscosity these four model systems of lipids are arranged as follows DOPC<DOPC/cholesterol<DHPC< DSPC from these it can be concluded that as the membrane environment becomes more rigid with higher viscosity the lifetime of TPE-Cy. Thus, encourages the model lipids to further discover the utility of TPE-Cy in sensing intracellular viscosity.

Viscosity measurement is not only important factor in live cell viscosity measurement but it also assist as an important indicator for food spoilage and important quality parameter in liquid drinks. So herein, Xu[59] and co-workers designed a new tetranitrile anthracene (TPAEQ) near infrared fluorophore for determining the viscosity through aggregation-induced emission (Fig. 34).

This probe is constructed based on electron donor-acceptor (D-A) comprising of triphenylamine methyl ether as donor and anthraquinone diamino nitrile as acceptor which exhibit ICT effect. In low viscous medium the triphenylamine and alkyl ether can rotate very easily without restriction thus there is no fluorescence enhancement. However, as the viscosity of the medium increases the intramolecular rotation of the molecule is inhibited

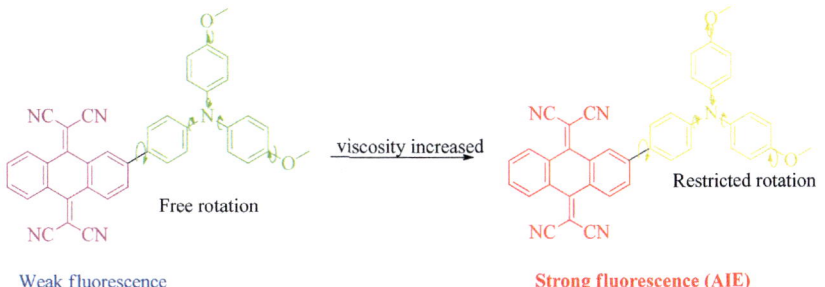

Fig. 34 Mechanistic pathway for detection of the probe TPAEQ for viscosity.

which results in enhancement in fluorescence. Fluorescence response of probe toward viscosity was measured in glycerol/water (0–99%) as shown in Fig. 35. However, it is experimentally observed that fluorescence intensity increases gradually as the viscosity of environment increases in the range of 1 cP (0% glycerol) to 956 cP (99% glycerol) due to which restriction of intramolecular rotation occurs. A linear relationship between fluorescence intensity and viscosity at 759 nm which is calculated by using Forster-Hoffmann equation. The viscosity sensitivity constant x was found to be 0.58 in water/glycerol. As most of the liquids are stored at lower temperature in order to reduce the spoilage. Thus the fluctuation of viscosity for glycerol at different temp. was studied. However, it was observed that there is no change observed for TPAEQ in presence of inorganic salts (NaCl, $CaCl_2$, $MgSO_4$, $AlCl_3$, $ZnCl_2$, $FeCl_3$, $NaNO_3$, K_2CO_3, Na_2SO_3, GSH, Cys, Hcy, Arg, Gly, disodium hydrogen phosphate, trisodium citrate dehydrate) but there is significant enhancement in fluorescence in presence of glycerol.

Furthermore, the probe can successfully applied in determining the fluid spoilage by viscosity measurement. The probe was utilized for sensing of fluid decay in different 8 different fresh liquid beverages such as fruit juice, Kaman juice, Jasmine tea juice, Pear juice, Sea buckthorn Juice, Jam, liquor, and milk (16.1 cP 3.5 cP 6.8 cP 2.2 cP 58.8 cP 112.2 cP 2.0 cP, 5.6 cP at 25 °C) respectively. In order to study the viscosity to all these fluids the food thickeners were added in an amount ranging from 1 to 10 g kg^{-1}. Generally sodium carboxymethyl cellulose pectin and xanthan gum are the most common food thickener used for analysis. Moreover fluorescence intensity increases as the concentration of thickeners increases.

Herein, a new class of two-photon AIE active styrylquinoline fluorophores HAPHs with excellent two photon characteristics were designed by Dou[60] and coworkers. These luminophores are developed by simple quinoline structure in simple steps (Fig. 36). Initially, strong fluorescence is observed in solid state for (E)-2-styrylquinoline (HAPH) but displays very weak fluorescence in organic solvents. However, the other fluorophore (E)-3-styrylquinoline (AQ-3S) and (E)-3-styrylquinoline AQ-6S demonstrated strong fluorescence in organic solvents hence suggests that (E)-2-styrylquinoline (AQ6S) displays the AIE characteristics. The chemical structures of the different styrylquinoline derivatives (Fig. 37). Interestingly it was observed that the synthesized HAPH-1 AIEgen showed good AIE characteristics. However, the styrylquinolines showed good AIE effect

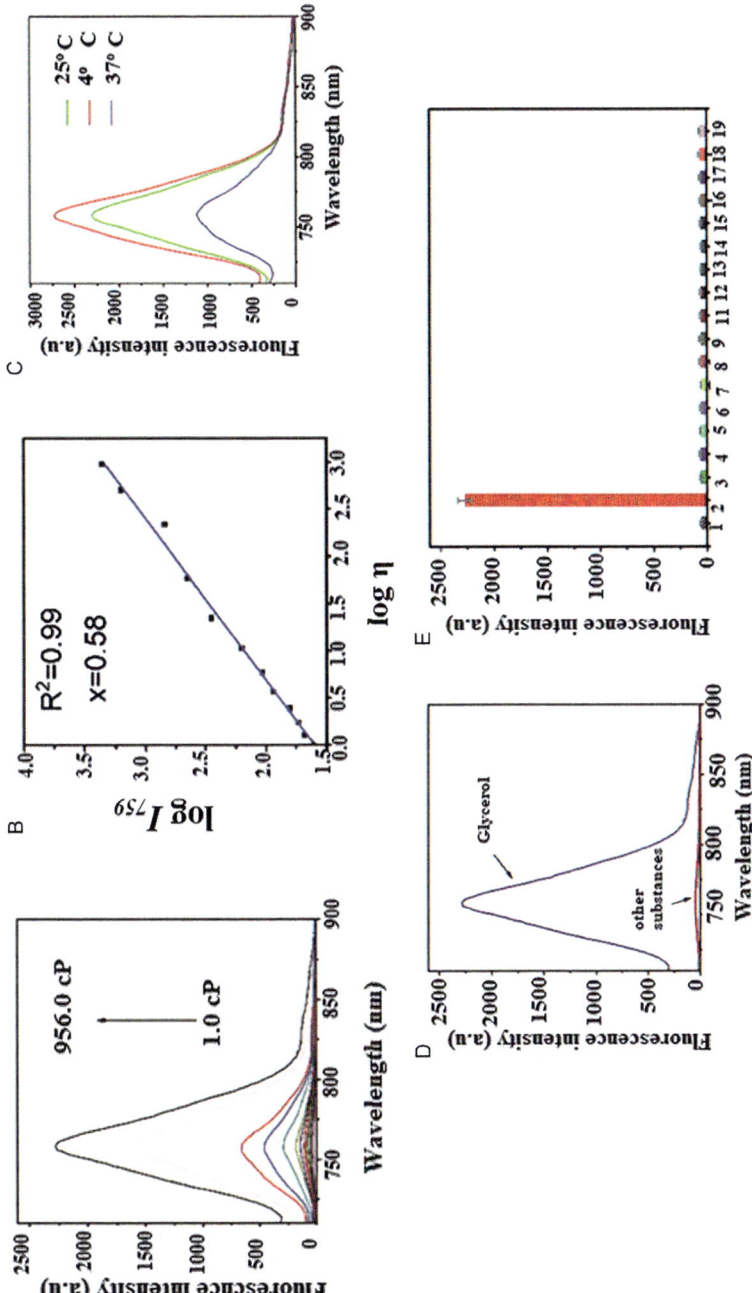

Fig. 35 (A) Illustration of PL spectra in water–glycerol fraction for TPAEQ probe in water–glycerol fraction with different fraction of glycerol (0–99%) at 650 nm. (B) The plot shows the linear relationship the same. (C) A plot illustrating fluorescence emission intensity recorded at different temperatures of TPAEQ (4 °C, 25 °C and 37 °C), (D) the PL spectra of TPAEQ in glycerol water fraction in (1% DMSO) and in presence of different metal salts. (E) PL intensity of TPAEQ with various substances (λex = 650 nm). *Reprinted from Xu L, Ni L, Zeng F, Wu S. Analyst. 2020;145:844–850 with permission of The Royal Society of Chemistry.*

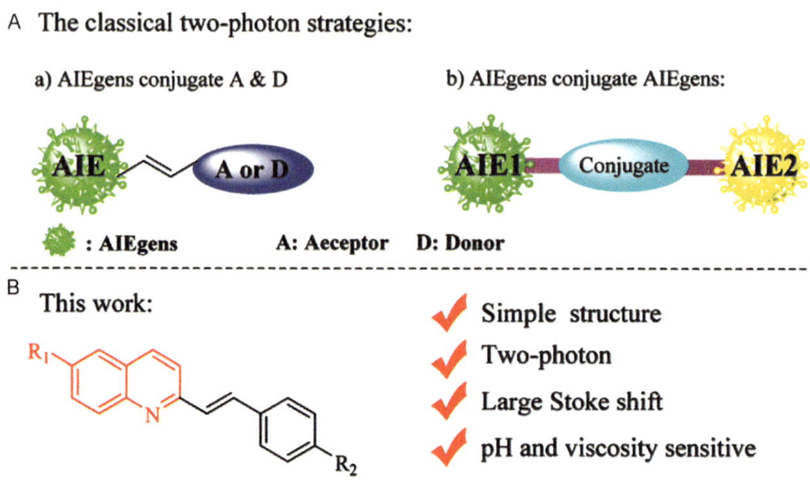

Fig. 36 Illustration of the styrylquinoline derivative (A) their current two photon design and (B) The structure of AIE luminogens (AIEgens). *Reprinted from Dou Y, Liu J, Zhang F, Cai C, Zhu Q, Kenry. J Mater Chem B. 2019;7:7771–7775 with permission of The Royal Society of Chemistry.*

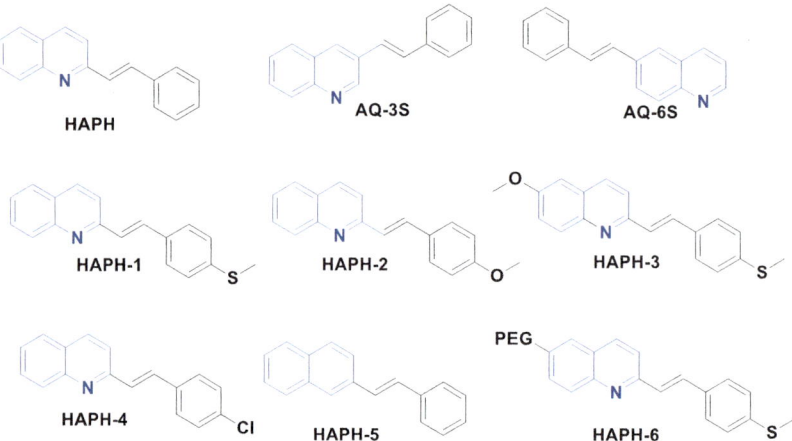

Fig. 37 Representation of different derivatives of HAPHs AIEgens.

when functionalized at C2 and C6 points (HAPH 1–3) with electron donor substitution at styryl ring. The fluorophore do not exhibit any AIE effect when HAPH-5 replaces quinoline with naphthalene.

Since, HAPH-1 show good AIE activity hence it is further used to study the biological application for pH and viscosity sensing. Moreover, HAPH-1 contains quinoline N which favors the hydrogen bonding under acidic

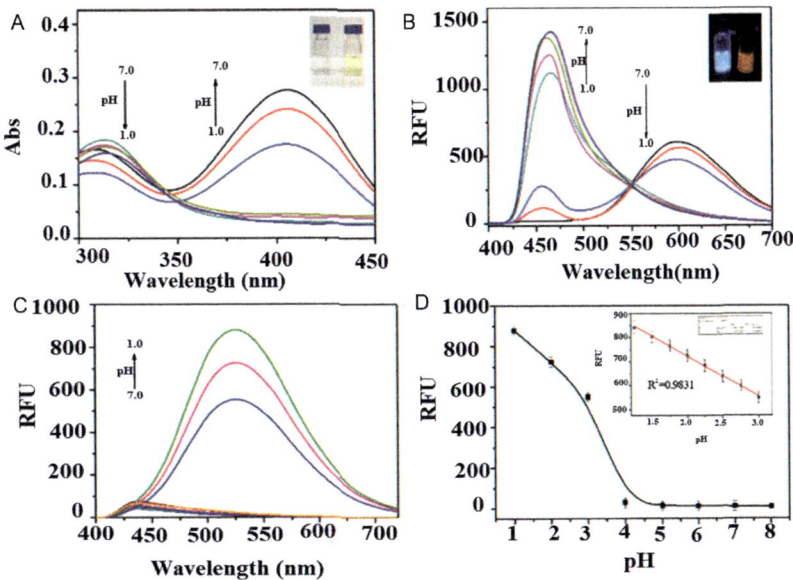

Fig. 38 (A) A UV–vis absorption spectral representation. (B) The PL spectra for HAPH-1 at 360 nm in pH range of 1–7 in water. (C) The plot of fluorescence intensity (580 nm) in the pH range of 1.00–7.00 at $\lambda_{ex}=410$ nm. (D) A graph of fluorescence intensity (580 nm) at $\lambda_{ex}=410$ nm and. Inset: A plot of fluorescence intensity vs pH in the low pH range of (1.25–3.00) with $R^2=0.9831$. *Reprinted from Dou Y, Liu J, Zhang F, Cai C, Zhu Q, Kenry. J Mater Chem B. 2019;7:7771–7775 with the permission of The Royal Society of Chemistry.*

condition hence it was studied for pH sensitivity (Fig. 38) The absorbance at 320 nm decreases as the pH decreases from 7.0 to 1.0 resulting in to appearance of new peak at 420 nm with increasing intensity. An intense emission band was observed at pH higher than 4.0 at 580 nm ($\lambda ex=410$ nm). However, no prominent AIE effect observed for the probe when pH is less than 4.0 According to previous studies TICT fluorophores can be utilized for viscosity sensing therefore the probe was studied for viscosity sensitive nature in different water and glycerol fractions. At 90% of glycerol fraction the viscosity of the solution increases from 1.01 cP to 950.17 cP. (Fig. 38A–C) The fluorescence response for the probe at 470 nm was observed with increasing emission at 550 nm and new absorption peak appeared at 360 nm upon addition of glycerol with a good linear relationship ($R^2=0.998$). Strong fluorescence was observed at 580 nm below pH 4 and due to enhancement in ICT effect emission intensity increases with increasing viscosity (Fig. 38D).

Because of its high applicability in pH and viscosity, the probe is considered a useful platform for biological application. Thus, HAPH-1 is an excellent probe for determination of intracellular viscosity.

5. Conclusion

Recently years, many researchers have progressively exploring use of the concept of aggregation induced emission (AIE) activity of developing various luminescent material for potential practical application to name few sensing, photoluminescence, solar cells, biology, medicine. Various research efforts by several groups have developed with various kinds of stimuli-responsive AIE active materials such as fluorescent organic molecules, nanoparticles, different molecular rotors, polymer-based nanogels these AIE active molecules can be used to study the responses to various environmental changes such as light, temperature, pressure, pH,[9] and viscosity. Among all these parameters temperature, pH and viscosity are the major factors responsible for biological activities occurring in the living system. This book chapter mostly insightful growth and progress in the field of sensing especially, temperature, pH, and viscosity sensing. Most of the subtopic focused on aggregation-induced emission (AIE) active molecules for measuring intracellular pH, viscosity, and temperature which is more described by using fluorescence and absorption study. A special emphasis is provided on AIE active fluorescent molecules, molecular rotors, polymeric nanomaterials which are considered as the important aspects of sense. In future, these AIE active molecules may useful for real world applications not only biological and medicinal sciences but also for material chemistry for other applications such as optical and colorimetric sensing and visualization of biomolecules and many more applications.

Acknowledgments

S.V.B. University Grant Commission (UGC) Faculty Research Program, New Delhi, India (F.4-5(50-FRP)(IV-Cycle)/2017(BSR)) for an award of Professorship and also acknowledges Council of Scientific & Industrial Research (CSIR), New Delhi code No. 02(0357)/19/EMR-II and DST-Goa for providing support.

References

1. Chen Y, Lam JWY, Kwok RTK, Liu B, Tang BZ. *Mater Horizons*. 2019;6:428–433.
2. Hong Y, Lam JWY, Tang BZ. *Chem Soc Rev*. 2011;40:5361–5388.
3. Mei J, Leung NLC, Kwok RTK, Lam JWY, Tang BZ. *Chem Rev*. 2015;115:11718–11940.

4. Hong Y, Lam JWY, Tang BZ. *Chem Commun.* 2009;4332–4353.
5. Liu J, Lam JWY, Tang BZ. *Chem Rev.* 2009;109:5799–5867.
6. Joglekar M, Trewyn BG. *Biotechnol J.* 2013;8:931–945.
7. Mazza MMA, Raymo FM. *J Mater Chem C.* 2019;7:5333–5342.
8. Shi P, Deng D, He C, et al. *Dye Pigment.* 2020;173, 107884.
9. Han J, Burgess K. *Chem Rev.* 2010;110:2709–2728.
10. Chen Z, Ding Z, Zhang G, Tian L, Zhang X. *Molecules.* 2018;23:1–12.
11. Qiao J, Chen C, Qi L, et al. *J Mater Chem B.* 2014;2:7544–7550.
12. Gao H, Kam C, Chou TY, Wu MY, Zhao X, Chen S. *Nanoscale Horizons.* 2020; 5:488–494.
13. Saha B, Ruidas B, Mete S, Mukhopadhyay CD, Bauri K, De P. *Chem Sci.* 2020;11:141–147.
14. Li T, He S, Qu J, et al. *J Mater Chem C.* 2016;4:2964–2970.
15. Wang Z, Yong TY, Wan J, et al. *ACS Appl Mater Interfaces.* 2015;7:3420–3425.
16. Yallapu MM, Jaggi M, Chauhan SC. *Drug Discov Today.* 2011;16:457–463.
17. Saunders BR, Laajam N, Daly E, Teow S, Hu X, Stepto R. *Adv Colloid Interface Sci.* 2009;147–148:251–262.
18. Chen CY, Chen CT. *Chem Commun.* 2011;47:994–996.
19. Zhao Y, Wu Y, Chen S, Deng H, Zhu X. *Macromolecules.* 2018;51:5234–5244.
20. Tang L, Jin JK, Qin A, et al. *Chem Commun.* 2009;4974–4976.
21. Zhou H, Liu F, Wang X, et al. *J Mater Chem C.* 2015;3:5490–5498.
22. Meng L, Jiang S, Song M, et al. *ACS Appl Mater Interfaces.* 2020;12:26842–26851.
23. Pandey UP, Thilagar P. *Adv Opt Mater.* 2020;8:1–16.
24. Matsuo K, Saito S, Yamaguchi S. *J Am Chem Soc.* 2014;136:12580–12583.
25. Feng J, Tian K, Hu D, et al. *Angew Chem Int Ed.* 2011;50:8072–8076.
26. Corrie SR, Coffey JW, Islam J, Markey KA, Kendall MAF. *Analyst.* 2015;140: 4350–4364.
27. Shamsipur M, Barati A, Nematifar Z. *J Photochem Photobiol C Photchem Rev.* 2019; 39:76–141.
28. Chen S, Liu J, Liu Y, et al. *Chem Sci.* 2012;3:1804–1809.
29. Chen S, Hong Y, Liu Y, et al. *J Am Chem Soc.* 2013;135:4926–4929.
30. Li K, Feng Q, Niu G, et al. *ACS Sensors.* 2018;3:920–928.
31. Lin N, Chen X, Yan S, et al. *RSC Adv.* 2016;6:25416–25419.
32. Wang J, Xia S, Bi J, et al. *Bioconjug Chem.* 2018;29:1406–1418.
33. Ma X, Cheng J, Liu J, Zhou X, Xiang H. *New J Chem.* 2015;39:492–500.
34. Rananaware A, Bhosale RS, Patil H, et al. *RSC Adv.* 2014;4:59078–59082.
35. Wang X, Wang H, Niu Y, Wang Y, Feng L. *Spectrochim Acta A Mol Biomol Spectrosc.* 2020;226:117650.
36. Qiu J, Jiang S, Guo H, Yang F. *Dye Pigment.* 2018;157:351–358.
37. Wang Z, Ye JH, Li J, Bai Y, Zhang W, He W. *RSC Adv.* 2015;5:8912–8917.
38. Yang DJ, Lin LY, Huang PC, Gao JY, Hong JL. *React Funct Polym.* 2016;108:47–53.
39. Zhao Y, Shi C, Yang X, et al. *ACS Nano.* 2016;10:5856–5863.
40. Jiang Y, Yang X, Ma C, et al. *ACS Appl Mater Interfaces.* 2014;6:4650–4657.
41. Feng Q, Li Y, Wang L, et al. *Chem Commun.* 2016;52:3123–3126.
42. Kuimova MK, Yahioglu G, Levitt JA, Suhling K. *J Am Chem Soc.* 2008;130:6672–6673.
43. Haidekker MA, Brady TP, Lichlyter D, Theodorakis EA. *J Am Chem Soc.* 2006; 128:398–399.
44. Stutts MJ, Canessa CM, Olsen JC, et al. *Sci Rep.* 1995;269:847–850.
45. Luby-Phelps K. *Int Rev Cytol.* 1999;192:189–221.
46. Harkness J. *Biorheology.* 1971;8:171–193.
47. Yang Z, Fan J, Peng X. *Funct Opt Imaging.* 2011;2011:6626–6635.
48. Haidekker MA, Theodorakis EA. *J Biol Eng.* 2010;4:1–14.

49. Sutharsan J, Lichlyter D, Wright NE, Dakanali M, Haidekker MA, Theodorakis EA. *Tetrahedron.* 2010;66:2582–2588.
50. Luby-Phelps K, Mujumdar S, Mujumdar RB, Ernst LA, Galbraith W, Waggoner AS. *Biophys J.* 1993;65:236–242.
51. Haidekker MA, Brady TP, Chalian SH, Akers W, Lichlyter D, Theodorakis EA. *Bioorg Chem.* 2004;32:274–289.
52. Wang M, Zhang Y, Yue X, Yao S, Bondar MV, Belfield KD. *Molecules.* 2016;21:1–12.
53. Kumbhar HS, Deshpande SS, Shankarling GS. *ChemistrySelect.* 2016;1:2058–2064.
54. Lee SC, Heo J, Woo HC, et al. *Chem Eur J.* 2018;24:13706–13718.
55. Gao M, Tang BZ. *ACS Sensors.* 2017;2:1382–1399.
56. Zhu YY, Wu XD, Gu SX, Pu L. *J Am Chem Soc.* 2019;141:175–181.
57. Yang Z, Huo Y, Liu Y, et al. *RSC Adv.* 2019;9:32219–32225.
58. Chen S, Hong Y, Zeng Y, et al. *Chem Eur J.* 2015;21:4315–4320.
59. Xu L, Ni L, Zeng F, Wu S. *Analyst.* 2020;145:844–850.
60. Dou Y, Liu J, Zhang F, Cai C, Zhu Q, Kenry. *J Mater Chem B.* 2019;7:7771–7775.

CHAPTER THREE

Advances in aggregation induced emission (AIE) materials in biosensing and imaging of bacteria

Mulaka Maruthi[a] and Suresh K. Kalangi[b],*

[a]Department of Biochemistry, Central University of Haryana, Mahendergarh, India
[b]Amity Stem cell Institute, Amity Medical School, Amity University Haryana, Amity Education Valley Pachgaon, Gurugram, India
*Corresponding author: e-mail address: skkalangi@ggn.amity.edu

Contents

1. Introduction	62
2. Aggregation induced materials	64
3. Photodynamic inactivation	65
4. Natural biocompatible AIE materials	66
5. Applications in biosensing	67
6. Potential application of AIEgens in imaging and killing of bacteria	69
6.1 Detection of bacterial viability	69
6.2 Single platform array to detect multiple bacterial species	70
6.3 pH dependent detection and clearance of microbes	72
6.4 Phage-guided AIE bioconjugates for imaging and killing of bacteria	73
7. Toxicology aspects of AIEgens	74
8. Future perspectives	75
Acknowledgments	75
References	75

Abstract

With their ubiquitous nature, bacteria have had a significant impact on human health and evolution. Though as commensals residing in/on our bodies several bacterial communities support our health in many ways, bacteria remain one of the major causes of infectious diseases that plague the human world. Adding to this, emergence of antibiotic resistant strains limited the use of available antibiotics. The current available techniques to prevent and control such infections remain insufficient. This has been proven during one of greatest pandemic of our generation, COVID-19. It has been observed that bacterial coinfections were predominantly observed in COVID-19 patients, despite antibiotic treatment. Such higher rates of coinfections in critical patients even after antibiotic treatment is a matter of concern. Owing to many reasons across the world drug resistance in bacteria is posing a major problem i. According to Center for Disease control (CDC) antibiotic report threats (AR), 2019 more than 2.8 million antibiotic resistant cases were reported, and more than 35,000 were dead among them in USA alone. In both normal and pandemic conditions, failure of identifying infectious agent has played a major role. This strongly prompts the need to improve upon the existing techniques to not just effective identification of an unknown bacterium, but also to discriminate normal Vs drug resistant strains. New techniques based on Aggregation Induced Emission (AIE) are not only simple and rapid but also have high accuracy to visualize infection and differentiate many strains of bacteria based on biomolecular variations which has been discussed in this chapter.

1. Introduction

Public health remains a major concern in developing and developed nations alike. Human race is threatened by different infections caused by pathogens like bacteria and fungi, which account for more than 2 million lives every year worldwide. In other way, intestinal bacteria play important roles, including food digestion, generation of essential micronutrients (e.g., Vitamin K) and contribute to the overall immunity in adults.[1] Contrarily, bacteria remain a major cause of fatal infectious diseases globally. Emerging antibiotic resistant bacterial strains further exaggerates this problem. In case of bacterial infections, diagnosis, and treatment at the early stage will be more effective to arrest the growth of diseases. Further, identification of causal organism is crucial, and is the foremost step, in clinical practice which helps in making informed decisions such as selection of drugs. Another area where bacterial identification remains a high priority is the food industry, for detection of food contamination. Unfortunately, emergence of drug-resistant bacterial pathogens, non-availability of suitable biomarkers and time-consuming screening of infections in patients pose a grave challenge to human health. Identification and classification of microbes with high efficiency and accuracy will facilitate rational use of antibiotics in clinical settings.

Recently, due to environmental pollution and climate change, the prevalence of water-borne pathogens has risen significantly.[2] For the detection of bacterial contamination, different methods are available. However, the currently available methods have some limitations such as longer incubation periods (e.g., standard plate count) and requirement for trained personnel (e.g., polymerase chain reaction). Additionally, development of methods for fast, reliable and instrument free detection of bacteria and other pathogens remains challenging. To overcome these limitations, there is an urgent need to develop whole bacterium analysis without involving complicated microbiological preparations. This not only provide greater understanding of the pathogen under investigation, but also, guide the development of novel therapeutic strategies. Theranostics is an emerging concept of combining diagnostic imaging and therapeutic intervention which is widely used in the treatment of bacterial diseases and systemic diseases like cancer.[3] Various radiological, fluorescence, and magnetic resonance imaging (MRI) techniques have been developed based on the specific interactions of ligand-receptor or antigen-antibody interactions[4] and are commonly used diagnostic techniques for more accuracy. Nevertheless, these are relatively expensive and have limited usage in real-time pathogen detection and tracking of tumor in intraoperative surgeries. In many instances, the variety of microbes to be detected may be novel and not detected by the probes with specific recognition moieties making the techniques ineffective. Thus, a simple light based turn on or off like rapid diagnosing method which can distinguish differentially expressed biomolecules on different strains of same microbe, are high on demand for efficient and accurate detection of microbes.

The classification and identification of bacteria and fungi relies on a labor intensive battery of tests such as Gram staining, PCR, genome sequencing, Raman Spectroscopy. These techniques are not only complicated, but also, require sophisticated equipment.[5] Additionally, integration of therapeutic agents is very much complicated process for cancer treatment, especially in coinfection cases.[3] Therefore, developing new strategies based on single component will largely minimize the design process of the targeted system for both imaging and killing of bacteria and cancer cells, and enhancing the theranostic purpose. Most bacterial surfaces are negatively charged, and are responsible for changes in the infected site microenvironment such as alterations in pH, and temperature. Additionally toxins and lipases released from bacteria also contribute to the unique characteristics of the infected site microenvironment.[6,7] Recently, the dependency on fluorimetry has become common in screening, and imaging applications due to high sensitivity and selectivity, easy to operate, noninvasive and rapid response.[8] In the recent past, a special class of luminogens with aggregation induced emission (AIEgens) characteristics has received global attention among researchers.[9–11] Further, AIEgens exhibit good biocompatibility, high quantum efficiency, and are photo stable. Thus, AIEgens are suitable for diverse applications such as biosensors, cell imaging, electroluminescent materials, and thus have great potential to use in biological and biomedical applications. In this chapter, we have reviewed the principles, properties, and applications of different AIE materials with respect to imaging or sensing of bacteria.

2. Aggregation induced materials

Individual molecules yet times end to aggregate into well-ordered structures supported by non-covalent interactions including hydrogen bonds, electrostatic interactions, π-π (pi-pi) interaction, hydrophobic interactions, and charge transfer effects.[12,13] Such aggregation a high concentrations is known to cause quenching of traditional fluorophores, a phenomenon called as Aggregation Caused Quenching (ACQ). The notorious ACQ effect has limited the use of luminogens at very dilute concentrations, which leads to serious photobleaching, effecting the imaging and other applications.[14] AIE is a photophysical phenomenon to monitor the light emitting processes from the aggregation of weakly or non-emissive luminogens.[11,15] In the solution, AIEgens are in monomeric form and remain non-emissive, but when aggregated restricted intra-molecular rotations (RIR)[10] cause proscription of energy dissipation resulting in high emission. The RIR restricts the energy dissipation of AIEgens through the nonradiative decay pathways, thus overcoming the ACQ effect and provide a way to develop efficient biochemicals.[16,17] Besides being biocompatible AIEgens have high quantum efficiency and photo-stability. These characteristics make them suitable for multiple applications in diverse fields such as therapeutics, biosensors, electroluminescent materials, bioimaging, optical devices etc.[11]

High scientific value and greater application potential for AIE phenomenon has attracted growing research efforts in various application fields. The development of AIEgens have opened new avenues for the applications in biosensing, chemotherapy, bioimaging and optoelectronics.[18] Additionally, some AIEgens have demonstrated production of reactive oxygen species (ROS) upon aggregation. This offers an unique opportunity toward development of light-up probes for image guided photodynamic therapy (PDT) for eliminating bacteria and cancer cells.[19–21] The fluorescence strength of AIEgens could be fine-tuned by using mixed solvent system containing organic water. Because most of the assays are carried out in aqueous buffer solutions, water solubility is a key requirement for the AIEgens along with the biocompatibility.[20] Among the AIEgens, owing to its simple synthesis and functionalization methods, Tetraphenylethene (TPE) is the smartest AIE-fluorogen reported so far to use in a variety of contexts[16] such

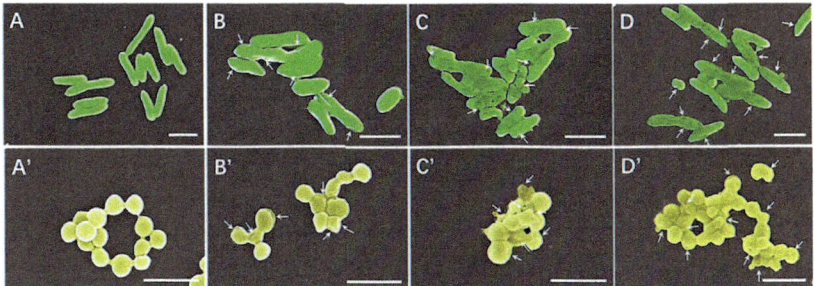

Fig. 1 Micromorphological changes of *E. coli* (A–D) and *S. aureus* (A′–D′) incubated with TPE-Cn Im compounds. The scanning electron microscopy (SEM) images shows the collapsed and distorted cell walls of the bacteria (indicated by arrows). Scale bar 2 μm. *Image from Shi J, Wang M, Sun Z, et al. Aggregation-induced emission-based ionic liquids for bacterial killing, imaging, cell labeling, and bacterial detection in blood cells.* Acta Biomater. *2019;97:247–259.*

as bioimaging, sensing, mechanochromism and as light harvesting material.[22,23] A variety of functionalized TPE-derivatives were reported for their use in bioassays and imaging of macromolecules in the organisms[24] (Fig. 1). A new promising strategy to eradicate pathogenic organisms have been developed using photosensitizers and AIEgens, this process is called photodynamic inactivation (PDI) or antimicrobial photodynamic therapy (PDT).

3. Photodynamic inactivation

Photodynamic inactivation (PDI) the phenomenon applied in photodynamic therapy (PDT) has become the recent promising tool for antibacterial activity and clearing cancer cells. PDI depends on light-sensitive, non-toxic photosensitizers (PS), which generate ROS under light irradiation.[25] The generated ROS can damage the bacterial cell walls by oxidation and subsequently destroy the bacteria. Photosensitizers are fluorogenic and could be utilized for image guided antibacterial functions. However, use photosensitizers require repeated washing steps which limits their use in real time detection of bacteria. Most of the photosensitizers used for PDI are hydrophobic, their interaction with bacterial cell wall causes aggregation resulting in fluorescence quenching and reduced ROS generation. Compromising the quality of imaging and PDT.[26] The effects of quenching were overcome by the recently developed AIEgens that could efficiently generate ROS under aggregated conditions. AIEgen based

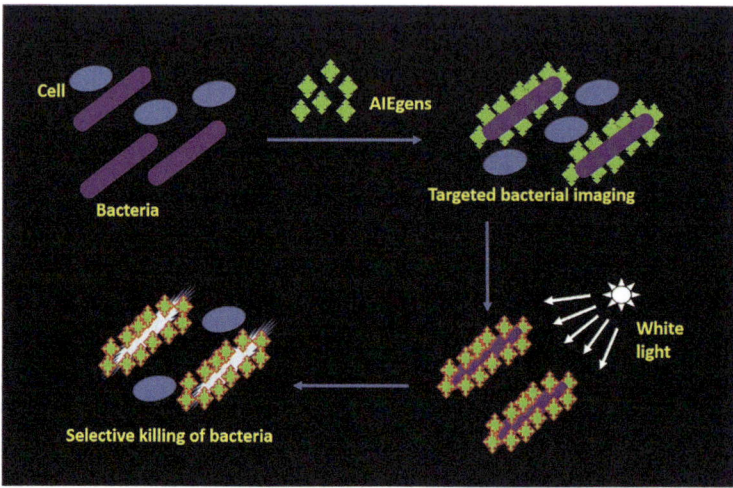

Fig. 2 The schematic illustration of AIEgens for selective targeting, imaging, and killing of bacteria. *Redrawn from Gao M, Hu Q, Feng G, et al. A multifunctional probe with aggregation-induced emission characteristics for selective fluorescence imaging and photodynamic killing of bacteria over mammalian cells. Adv Healthc Mater. 2015;4 (5):659–663.*

photosensitizers therefore would revolutionize the development of light up probes utilized in image guided PDT[8] (Fig. 2).

Being PDI targets are mostly external components of the bacteria, it does not require the photosensitizers to enter the bacterial cytosol all the time unlike the antimicrobial agents,[27] therefore bacteria cannot effectively develop resistance against AIEgens used in PDI. So far, several cationic AIEgens have been shown the activity of killing bacteria based on electrostatic interactions. To give a positive charge to the AIEgens, ammonium slats and Zinc (II)-dipicolylamine (ZnDPA) are often employed in a coordinated complex, among these ZnDPA is the most popularly used, it has higher positive charge and has stronger binding affinity for bacteria.[28,29] Elevated ROS generation was reported with some AIEgens, which may be effectively used for light induced cancer cell ablation.[30]

4. Natural biocompatible AIE materials

The most abundant natural polysaccharides including starch, cellulose and chitosan have been studied extensively in the industrial material applications and biomedicine. The fluorescent properties of biomacromolecules have received very little attention. The versatility, and biocompatibility of biomacromolecules will aid the development of new generation of biosensors with potential applications in a wide range of domains including diagnostics.[31–33] Chitosan, a product of chitin deacetylation, is abundant in nature. Chitosan is a non-toxic bioactive polymer with bactericidal property. Further, chitosan is easy to modify, and biodegradable. All these factors render chitosan to be used in a wide variety of fields including food and water treatment, and medicine.[34,35] But, the AIE effect of chitosan has not been reported without chemical modification.

Recent studies have made attempts to utilize chitosan as AIEgen in bacterial imaging, and metal ion concentration detection.[36] The fluorogenic activity of chitosan mainly depends on two aspects, modification of the fluorescent group and crosslinking with aldehyde groups.[34,37] The Schiff base intermediate formed during the crosslinking of aldehyde to chitosan causes autofluorescence due to π-π transition of the CN bond. This has provided a possibility of using chitosan as a AIE material.[36] It was also found that chitosan emits a range of colors from blue to red in an excitation wavelength dependent manner, offering excellent opportunities for cellular imaging. Moreover, chitosan is capable of efficiently detecting Fe^{3+}, making it a potent material for the detection of Fe^{3+} concentration in the environment and inside the living systems. Additionally, chitosan was used to detect *E. coli* by aggregation of chitosan on the *E. coli* surface, which makes chitosan a better option for real-time image guided monitoring of antibacterial processes.[36] Together, Chitosan offers a novel platform for multifunctional applications.

5. Applications in biosensing

Gram's staining is the first and gold standard to discriminate bacteria from all microbes and on broad, which is simple staining method based on bacterial surface changes. With the emerging bacterial diversities and surface adaptations, it became a tedious task to use suitable technique identify the bacterial species, which further laid the foundation for the use of targeted and sensitive fluorescence based probes in diagnosis of the disease. Qualitative and quantitative staining of live bacteria and imaging the intra-cellular bacteria, distinctly which are viable but not culturable in laboratory have attained main focus in the era of high-resolution microscopes to aid the diagnosis and treatments.[38] In contrast to the time-consuming methods, rapid, instantly staining and live imaging/diagnostic methods have many advantageous in monitoring sudden changes in environment, pharmaceutical developments, food processing, biomedical research and clinical diagnosis aspects. Faster, economic, and sensitive diagnosis in resource limiting conditions demand a simpler way like color or light turn on/off reactions as a read out, just by adding a chemical compound at room temperatures. AIEgens are unique probes that offer highly sensitive detection process rapidly by turning off/on light upon binding to the target. AIEgens are tunable, targetable to modulate according to the requirement.

Rapid detection of bacteria is valuable in different fields such as clinical diagnosis, pharmaceutical development, food processing and environmental sciences.[39,40] In all these applications, viability of the bacteria is an important aspect which offers significant improvement for simple, rapid, and accurate detection. The methods involved in the detection and identification of bacteria such as standard plate count method, PCR and high-resolution microscopy are time consuming and laborious processes. Moreover, these techniques are based on the indirect recognition of alterations in bacterial morphology and membrane potential.[41,42]

In this regard, fluorescence-based methods may offer direct method for real-time bacterial detection and identification, without requiring sophisticated instrumentation.[43,44] For this purpose, the most essential criterion for achieving rapid detection of bacteria using fluorescence methods is the development of fluorescent dyes with high sensitivity and specificity. Fluorescent detection of wild type bacteria could be readily achieved by using positively charged dyes that electrostatically bind to negatively charged bacterial surfaces. However, binding of multiple charged fluorescent dyes may cause severe alterations in bacterial states through induction of metabolic changes or by aggregation of the bacteria. Additionally, electrostatic adsorption based fluorescent labelling cannot distinguish between live and dead bacterial cells. Further, in contrast to small molecule antibiotics, macromolecular antibacterial compounds exhibit sustained bacterial inhibition with broad specificity. Also, such macromolecules are easy to be functionalized through covalent or non-covalent approaches.[45] In line with this, TPE core containing amphiphilic/cationic arms (TPE-star-P (DMA-co-BMA-co-Gd)) have been shown to exhibit fluorescence emission upon binding to bacteria. In addition, these molecules were reported to have anti-bacterial properties against both gram-positive and gram-negative strains.[21]

6. Potential application of AIEgens in imaging and killing of bacteria

Ubiquitous bacteria are intricately associated with every aspect of human ordinary life, in particular public health and human welfare.[46] Globally, millions of people are affected by bacterial infections every year.[47] To overcome the tragedies, several antibiotics have been developed for treating the infections and prevention of pathogens.[48] Inappropriate use of antibiotics is the leading cause for the rapid emergence of antibiotic resistance, which further increased the need for novel antibiotics.[47] However, antibiotic development is an expensive and time-consuming process. Considering these facts, real-time visualization and elimination of bacteria that cause infections and contaminations is an important task necessary for the human health and food security.[49]

6.1 Detection of bacterial viability

To date, various approaches have been established for the imaging and identification of bacteria. Among them, the methods based on fluorescent probes have been widely used because of their easy operation and high sensitivity.[50,51] For any fluorescence based rapid detection, the key is to be highly responsive and bacterial specific dye. In common, a counter ionic fluorescent dye is used to the charge on bacterial surface. But, normal fluorescent dyes, have been known for inducing metabolic changes and bacteria cell aggregation. Also, these dyes cannot distinguish between live and dead bacteria because of their electrostatic adsorption to the surface.

Most bacterial species contain abundant polysaccharides and peptidoglycan on their surfaces and cell wall.[52] Peptidoglycan accounts for 50–80% of the dry weight of gram-positive bacterial cells and 5–20% of gram-negative cells. The multiple hydroxyl groups of the surface polysaccharides present in various configurations are used as binding sites for fluorescent compounds.[52] The mechanism of specific binding of phenylboronic acids to the diols of the carbohydrates[53] has been adapted to develop phenylboronic acids conjugated to AIEgens resulting in the development of efficient indicator compounds such as TBE-2BA for the detection of polysaccharide surfaces of the bacteria.[23,54] TPE-2BA emitted fluorescence only when aggregated on binding to the cis-diols of D-glucose representing binding induced aggregation, which demonstrated the feasibility of specific imaging and detection of predetermined targets such as detecting live bacteria.[24] Using similar approach, hybrid systems containing b-galactosidase and cationic gold nanoparticles were fabricated for bacterial detection.[55] Also, a FRET system using quaternary ammonium functional group conjugated polymer was developed to probe the antimicrobial susceptibility, and to screen for potent antimicrobial compounds.[56] Thus, competitive supramolecular interactions between negatively charged bacterial surface and synthetic materials could be extensively utilized for sensing bacteria.[57]

To detect live bacteria, chemical specificity based methods are essential like targeting polysaccharides, or receptors on bacteria unlike whole surfaces. Bacterial surface polysaccharides possess several hydroxyl groups in different configurations and thus could be used as fluorescent binding sites. For instance phenylboronic acid specifically binds to diols present on carbohydrates.[58] This affinity inspired the design of biosensors to specifically target carbohydrates on bacteria. Upon conjugating phenylboronic acid to an unique indicator molecule, 4,4′-(1,2-diphenylethene-1,2-diyl)-bis-(4,1-phenylene) diboronic acid (TPE-2BA), emitted fluorescence only when bound to cis-diols of D-glucose to form aggregates. Tang and coworkers, in 2001, for the first time, have explained this phenomenon as AIE.[59]

Same Tang group in 2018, have successfully designed and demonstrated that, a morpholine-containing 2-(diphenylmethylene) hydrazono methyl naphthalene (DPAN), a derivative of M1-DPAN could specifically discriminate gram-positive bacteria from other bacteria, and when bound, it could yield 24 h long fluorescence signal which allowed to trace their infectivity toward mammalian cells.[60] DPAN-based AIEgens are not only applied to discriminate and visualize live pathogens, but also, provided platform to develop better tracers that could be used in real-time monitoring of infection dynamics under different treatments as shown in Fig. 3.

6.2 Single platform array to detect multiple bacterial species

AIEgens are used for the bacterial imaging. For instance, vancomycin based light up probe (a red emissive fluorescence probe), AIE-2Van, for specific recognition, and image guided photodynamic elimination of Gram-positive bacteria.[61] For visualization and elimination of both gram positive and gram-negative bacteria, TPE derived cationic amphiphilic molecules with ammonium groups are used which uses electrostatic interactions for binding.[62] In another approach, for developing wash free detection tools for

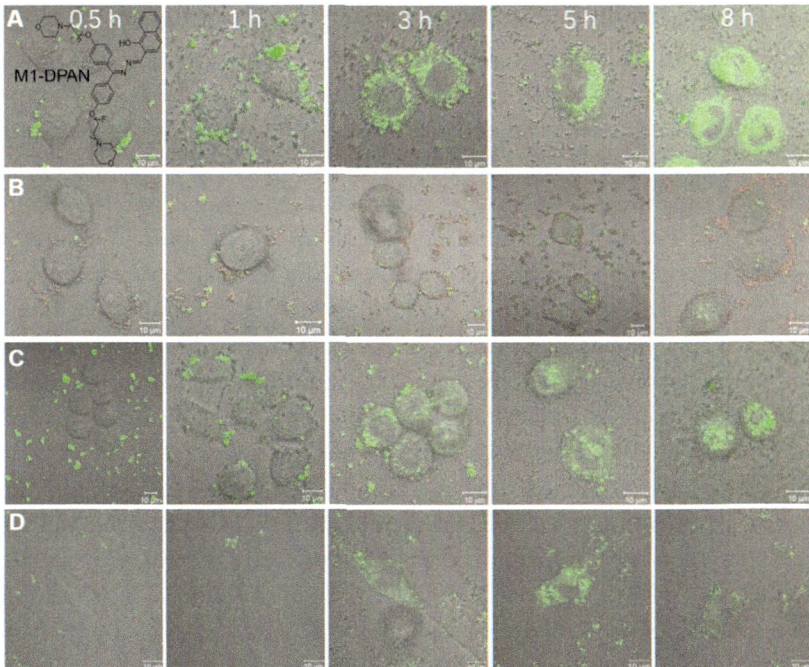

Fig. 3 The CLSM images of HeLa cells infected by M1-DPAN-labeled (A) living *S. aureus*, (B) *S. aureus* treated by 75% EtOH and (C) *S. aureus* treated with cephalothin and PI. (D) The CLSM images of NIH 3T3 cells infected by M1-DPAN-labeled living *S. aureus*. Reproduced with permission from Hu R, Zhou F, Zhou T, et al. Specific discrimination of gram-positive bacteria and direct visualization of its infection towards mammalian cells by a DPAN-based AIEgen. Biomaterials 2018;187:47–54. Copyright 2018 Elsevier.

imaging of bacteria, the abundance of lipoteichoic acid (LTA) on the cell walls of gram-positive bacteria have been exploited. LTA, an amphiphilic molecule with NH_3+ containing backbone offers a scope to develop an array of negatively charged AIE-probes as potential turn-on fluorescent molecules for effective imaging.[63]

An array of TPE based AIE materials with positive, negative, or neutral charges were synthesized with capability for broad spectrum bacterial imaging which were tested successfully on different gram-positive and gram-negative bacterial species.[61] The array of TPEs provided differential fluorescent responses due to multivalent nonspecific interactions between the bacterial cell surface molecules and TPEs provided an excellent platform for bacterial discrimination without the use of radioactive markers and antibodies.[61]

6.3 pH dependent detection and clearance of microbes

Generally, synthesized chemical compounds exhibit antimicrobial activity by stress-induced cell death which is dependent on the binding of the compounds to the cell wall receptors.[64] The same mechanism has been exploited to develop KB1, active site's OH group of KB1 might interact with the cell wall or plasma membrane proteins, upon adherence it will increase the cell membrane permeability by interacting with membrane proteins and other microbial molecules.[65] The increased permeability will lead to the loss of intracellular electrolytes such as Na+, Ca2+ and K+ resulting in the cell death.[65] In other instances, decrease in pH occurs at the site of infection due to local acidosis caused by the infiltration of macrophages and neutrophils during infection induced inflammatory response.[66,67] The usage of pH responsive agents with enhanced anti-bacterial activity can efficiently control infections with minimal side effects and high therapeutic efficacy.[67] The local acidosis principle has been adapted by Yang *et al.* to fabricate pH sensitive polymers such as quaternary pyridinium with on demand antimicrobial activity.[68] However, the potential toxicity associated with pyridinium units limit their use for biomedical applications. As an alternative poly (vinyl alcohol) (PVA), which has excellent biocompatibility and bio adhesive properties can be used as precursor to develop alternative pH sensitive polymeric materials for bacterial killing.[69]

Various fluorescent probe-based approaches are available for bacterial imaging and killing such as molecule-, polymer- and nanomaterial-based probes, which have been intensively used because of their easy handling and high sensitivity.[70] The AIE active materials (AIEgens) are of great scientific value and practical implications. AIEgens with light induced ROS generation capability have been applied to light enhanced bacterial elimination during photodynamic therapy.[46,71] A small AIEgen molecule TPE-Bac is an effective bacterial imaging and antibacterial material. Owing to their positively charged surfaces, these molecules were successfully able to penetrate the plasma membranes.[62] Another molecule, AIE-active DBPE polymerized to DBPE-DBO possess multiple positive charges on the polymers which provide multiple interaction points with bacterial cell surface. The hydrophobic alkyl chains of the polymers intercalate into the membrane components of the bacterial cell surface.[72] Thus, synergistically increasing the membrane permeability for the AIE active polymer DBPE-DBO into the bacteria. Additionally, DBPE-DBO could distort phospholipid arrangement on bacterial membranes which can increase bacterial toxicity. DBPE-DBO exhibited excellent photosensitizing properties generating

ROS (singlet oxygen), which exerted potential antibacterial activity after irradiating at room light conditions suggesting DBPE-DBO as an excellent candidate for PDT.[72]

6.4 Phage-guided AIE bioconjugates for imaging and killing of bacteria

Meanwhile, in the last year, i.e., 2020, Tang and coworkers came up with much curious working principle, where AIE bioconjugates can be designed and utilized through the guidance of bacteria phage, which has natural capability to selectively target particular bacteria type. This group has showed that these conjugates are not only useful in tracing bacterium, but also kill them by using their exceptional capability of photodynamic therapeutic action. This novel strategy is based on integrating AIEgens like TVP-S with bacteriophage targeting *P. aeruginosa*. These molecules form a new class of antimicrobial bioconjugates (TVP-PAP) with AIE properties and excellent PDI activity.[73] AIEgens have been conjugated through an amino carboxyl reaction, without affecting the properties of both AIEgen and phage, thus allowing the real time monitoring of phage-bacterium interaction with high specificity as shown in Fig. 4.[73]

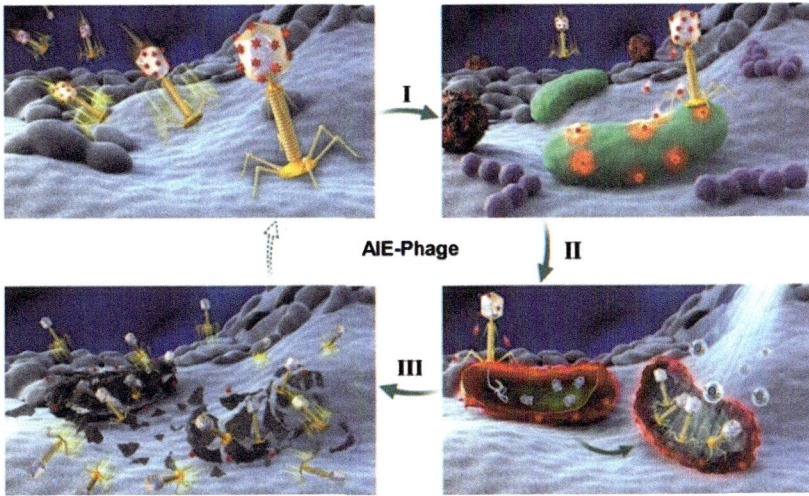

Fig. 4 Pictures depict the mechanism of phase guided bacterial targeting by discriminative imaging and synergetic killing through AIE bioconjugates. The phage guided targeting and killing include: (I) Identification of bacteria by AIEgen conjugated specific bacteriophage; (II) lighting up and imaging of the targeted bacteria and infectious phage by AIE generated fluorescence; and (III) the synergistic killing of the targeted bacteria by phage infections and AIE-base photodynamic inactivation.[73]. *Reprinted with permission from publishers.*

Together, phage conjugated AIE-based photodynamic inactivation pave a wonderful exciting platform for killing bacteria which has benefits both fluorescence properties, and efficient $_1O^2$ generation capability. TVP-PAP AIEgens showed good visualization and selective bacterial elimination properties, with minimal or no effects on other bacteria and normal mammalian cells as reported by Tang group. The in vitro test provided the evidence that even multidrug resistant *P. aeruginosa* could be killed with near 100% efficiency. This could be attributed to the synergistic effect within phage guided AIEgen, i.e., TVP-PAP.

7. Toxicology aspects of AIEgens

Though the AIE materials offer color tunability and structural diversity, the sources of AIEgens are organic synthesis with non-planar conjugated molecular confirmations and thus have poor water solubility making them environmentally disadvantageous because of the toxicity and poor degradation.[74–76] AIEgens are generally non emissive and non-toxic in dark but upon aggregation can produce toxic ROS, the ROS destroys the bacteria by oxidizing the bacterial cell walls. This unique feature provides a better chance to develop novel light up probes for image guided PDI of bacteria.[77] There is an urgent need to develop new class of AIEgens with good water solubility, degradation, and biocompatibility to reduce environmental toxicity.

The AIEgens can be effectively used for killing bacteria, as AIEgens such as TPETH-2Zn, AIE-ZnDPA and TPE-amphiphiles selectively images and exerts phototoxicity to bacteria over mammalian cells, this has been evident from the mammalian cell viability analysis using MTT assay on HeLa cells and Jurkat T cells. No significant mammalian cell cytotoxicity and binding by TPETH-2Zn to the mammalian cells was observed, whereas the bacterial killing efficiency was more than 95% at the same concentration.[61,78,79] This could be due to the selective binding affinities of the AIEgens toward bacteria. Moreover TPETH-2Zn is very selective toward targeted pathogenic bacteria than the nonpathogenic bacteria present in the solution for photoinactivation.[80] In vivo toxicity analysis has revealed that TPE based AIEgens have exhibited no or negligible toxicity.[81]

Certain copolymers have showed relatively high cytotoxicity against human red blood cells (RBCs) making them unfavorable for their potent applications. But, after quaternization reactions the cytotoxicity to RBCs has decreased drastically assessed by hemoglobin release assay, and these copolymers possess pH independent water solubility to accomplish bacterial detection under physiological conditions.[21] The relative low hemolysis rate and cell toxicity of quarternized copolymers of TPE-C1 allow them to act as light-up sensors for the visualization and elimination of pathogenic bacteria in blood cells.[21,82]

8. Future perspectives

Identification of bacteria and diseased cells plays crucial roles in clinical practice, which helps in the understanding of origin of the disease. A new class of AIEgens with good photostability, sensitivity and improved signal to noise ratio, are being used for various bioanalytical techniques, biosensing and cell biology applications. The studies also validate the capabilities of the AIEgens to differentiate between Gram-positive and Gram-negative bacteria, and cancer cells from normal cells.[81,83] The AIEgens can be further developed for comprehensive cancer and infectious disease therapies, with an approach to design multifunctional systems with accurate in vivo localization, diagnosis and killing of cancer cells and pathogenic bacteria.

The traditional AIEgens have the excitation wavelengths in the UV range with poor tissue penetrations.[80,81,84] This limitation can be overcome by the development of red and near-infrared emissive AIEgens. Such molecules benefit the study of in vivo localization of bacteria and cancer cells, and treatment. The anti-bacterial activity of the AIEgens can be enhanced by developing easy synthetic methods for generating fluorophores coupled with antimicrobial agents which will revolutionize the field of nano antibiotics.

Acknowledgments

Authors express their heartfelt thanks for the editors for choosing and the kind invitation toward the participation in this book writing.

References

1. Steed AL, Christophi GP, Kaiko GE, et al. The microbial metabolite desaminotyrosine protects from influenza through type I interferon. *Science*. 2017;357(6350):498–502.
2. Parks T, Hill AV, Chapman SJ. The perpetual challenge of infectious diseases. *N Engl J Med*. 2012;367(1):90 [author reply 90].
3. Kelkar SS, Reineke TM. Theranostics: combining imaging and therapy. *Bioconjug Chem*. 2011;22(10):1879–1903.
4. Budin G, Chung HJ, Lee H, Weissleder R. A magnetic gram stain for bacterial detection. *Angew Chem Int Ed Engl*. 2012;51(31):7752–7755.

5. Reisner BS, Woods GL. Times to detection of bacteria and yeasts in BACTEC 9240 blood culture bottles. *J Clin Microbiol*. 1999;37(6):2024–2026.
6. Sawyer RG, Spengler MD, Adams RB, Pruett TL. The peritoneal environment during infection. The effect of monomicrobial and polymicrobial bacteria on pO2 and pH. *Ann Surg*. 1991;213(3):253–260.
7. Xiong MH, Bao Y, Yang XZ, Wang YC, Sun B, Wang J. Lipase-sensitive polymeric triple-layered nanogel for "on-demand" drug delivery. *J Am Chem Soc*. 2012;134 (9):4355–4362.
8. Yuan Y, Feng G, Qin W, Tang BZ, Liu B. Targeted and image-guided photodynamic cancer therapy based on organic nanoparticles with aggregation-induced emission characteristics. *Chem Commun (Camb)*. 2014;50(63):8757–8760.
9. Gao M, Tang BZ. Fluorescent sensors based on aggregation-induced emission: recent advances and perspectives. *ACS Sens*. 2017;2(10):1382–1399.
10. Hong Y, Lam JW, Tang BZ. Aggregation-induced emission: phenomenon, mechanism and applications. *Chem Commun (Camb)*. 2009;(29):4332–4353.
11. Mei J, Leung NL, Kwok RT, Lam JW, Tang BZ. Aggregation-induced emission: together we shine, united we soar! *Chem Rev*. 2015;115(21):11718–11940.
12. Hartgerink JD, Beniash E, Stupp SI. Self-assembly and mineralization of peptide-amphiphile nanofibers. *Science*. 2001;294(5547):1684–1688.
13. Bhattacharya S, Samanta SK. Soft-nanocomposites of nanoparticles and nanocarbons with supramolecular and polymer gels and their applications. *Chem Rev*. 2016;116 (19):11967–12028.
14. Resch-Genger U, Grabolle M, Cavaliere-Jaricot S, Nitschke R, Nann T. Quantum dots versus organic dyes as fluorescent labels. *Nat Methods*. 2008;5(9):763–775.
15. Leung ACS, Zhao E, Kwok RTK, et al. An AIE-based bioprobe for differentiating the early and late stages of apoptosis mediated by H2O2. *J Mater Chem B*. 2016;4 (33):5510–5514.
16. Tong H, Hong Y, Dong Y, et al. Protein detection and quantitation by tetraphenylethene-based fluorescent probes with aggregation-induced emission characteristics. *J Phys Chem B*. 2007;111(40):11817–11823.
17. Liao G, Yao W, Zuo J. Preparation and characterization of zeolite/TiO(2) cement-based composites with excellent photocatalytic performance. *Materials (Basel)*. 2018;11 (12):2485.
18. Kim HN, Ren WX, Kim JS, Yoon J. Fluorescent and colorimetric sensors for detection of lead, cadmium, and mercury ions. *Chem Soc Rev*. 2012;41(8):3210–3244.
19. Feng G, Yuan Y, Fang H, et al. A light-up probe with aggregation-induced emission characteristics (AIE) for selective imaging, naked-eye detection and photodynamic killing of gram-positive bacteria. *Chem Commun (Camb)*. 2015;51(62):12490–12493.
20. Chen W, Li Q, Zheng W, et al. Identification of bacteria in water by a fluorescent array. *Angew Chem Int Ed Engl*. 2014;53(50):13734–13739.
21. Li Y, Yu H, Qian Y, Hu J, Liu S. Amphiphilic star copolymer-based bimodal fluorogenic/magnetic resonance probes for concomitant bacteria detection and inhibition. *Adv Mater*. 2014;26(39):6734–6741.
22. La DD, Bhosale SV, Jones LA, Bhosale SV. Tetraphenylethylene-based AIE-active probes for sensing applications. *ACS Appl Mater Interfaces*. 2018;10(15):12189–12216.
23. Liu Y, Deng C, Tang L, et al. Specific detection of D-glucose by a tetraphenylethene-based fluorescent sensor. *J Am Chem Soc*. 2011;133(4):660–663.
24. Kong TT, Zhao Z, Li Y, Wu F, Jin T, Tang BZ. Detecting live bacteria instantly utilizing AIE strategies. *J Mater Chem B*. 2018;6(37):5986–5991.
25. Shrestha A, Kishen A. Polycationic chitosan-conjugated photosensitizer for antibacterial photodynamic therapy. *Photochem Photobiol*. 2012;88(3):577–583.

26. Park SY, Baik HJ, Oh YT, Oh KT, Youn YS, Lee ES. A smart polysaccharide/drug conjugate for photodynamic therapy. *Angew Chem Int Ed Engl.* 2011;50(7):1644–1647.
27. Almeida A, Faustino MA, Tome JP. Photodynamic inactivation of bacteria: finding the effective targets. *Future Med Chem.* 2015;7(10):1221–1224.
28. Yuan Y, Zhang CJ, Gao M, Zhang R, Tang BZ, Liu B. Specific light-up bioprobe with aggregation-induced emission and activatable photoactivity for the targeted and image-guided photodynamic ablation of cancer cells. *Angew Chem Int Ed Engl.* 2015;54(6):1780–1786.
29. Zhao E, Chen Y, Chen S, et al. A luminogen with aggregation-induced emission characteristics for wash-free bacterial imaging, high-throughput antibiotics screening and bacterial susceptibility evaluation. *Adv Mater.* 2015;27(33):4931–4937.
30. Maisch T, Bosl C, Szeimies RM, Lehn N, Abels C. Photodynamic effects of novel XF porphyrin derivatives on prokaryotic and eukaryotic cells. *Antimicrob Agents Chemother.* 2005;49(4):1542–1552.
31. LogithKumar R, KeshavNarayan A, Dhivya S, Chawla A, Saravanan S, Selvamurugan N. A review of chitosan and its derivatives in bone tissue engineering. *Carbohydr Polym.* 2016;151:172–188.
32. Olsson RT, Azizi Samir MA, Salazar-Alvarez G, et al. Making flexible magnetic aerogels and stiff magnetic nanopaper using cellulose nanofibrils as templates. *Nat Nanotechnol.* 2010;5(8):584–588.
33. Dou X, Zhou Q, Chen X, et al. Clustering-triggered emission and persistent room temperature phosphorescence of sodium alginate. *Biomacromolecules.* 2018;19(6):2014–2022.
34. Younes I, Rinaudo M. Chitin and chitosan preparation from marine sources. Structure, properties and applications. *Mar Drugs.* 2015;13(3):1133–1174.
35. Liu GJ, Tian SN, Li CY, Xing GW, Zhou L. Aggregation-induced-emission materials with different electric charges as an artificial tongue: design, construction, and assembly with various pathogenic bacteria for effective bacterial imaging and discrimination. *ACS Appl Mater Interfaces.* 2017;9(34):28331–28338.
36. Dong Z, Cui H, Wang Y, Wang C, Li Y, Wang C. Biocompatible AIE material from natural resources: chitosan and its multifunctional applications. *Carbohydr Polym.* 2020;227:115338.
37. Wang K, Yuan X, Guo Z, Xu J, Chen Y. Red emissive cross-linked chitosan and their nanoparticles for imaging the nucleoli of living cells. *Carbohydr Polym.* 2014;102:699–707.
38. Camesano TA, Natan MJ, Logan BE. Observation of changes in bacterial cell morphology using tapping mode atomic force microscopy. *Langmuir.* 2000;16(10):4563–4572.
39. Oukacine F, Quirino JP, Garrelly L, Romestand B, Zou T, Cottet H. Simultaneous electrokinetic and hydrodynamic injection for high sensitivity bacteria analysis in capillary electrophoresis. *Anal Chem.* 2011;83(12):4949–4954.
40. Cheng MS, Lau SH, Chow VT, Toh CS. Membrane-based electrochemical nanobiosensor for *Escherichia coli* detection and analysis of cells viability. *Environ Sci Technol.* 2011;45(15):6453–6459.
41. Abdel-Hamid I, Ivnitski D, Atanasov P, Wilkins E. Flow-through immunofiltration assay system for rapid detection of *E. coli* O157:H7. *Biosens Bioelectron.* 1999;14(3):309–316.
42. Labib M, Zamay AS, Muharemagic D, Chechik AV, Bell JC, Berezovski MV. Aptamer-based viability impedimetric sensor for viruses. *Anal Chem.* 2012;84(4):1813–1816.
43. Strauber H, Muller S. Viability states of bacteria—specific mechanisms of selected probes. *Cytometry A.* 2010;77(7):623–634.
44. Keer JT, Birch L. Molecular methods for the assessment of bacterial viability. *J Microbiol Methods.* 2003;53(2):175–183.

45. Shao Q, Xing B. Enzyme responsive luminescent ruthenium(II) cephalosporin probe for intracellular imaging and photoinactivation of antibiotics resistant bacteria. *Chem Commun (Camb)*. 2012;48(12):1739–1741.
46. Kohanski MA, Dwyer DJ, Collins JJ. How antibiotics kill bacteria: from targets to networks. *Nat Rev Microbiol*. 2010;8(6):423–435.
47. Costerton JW, Stewart PS, Greenberg EP. Bacterial biofilms: a common cause of persistent infections. *Science*. 1999;284(5418):1318–1322.
48. Blaskovich MAT, Hansford KA, Butler MS, Jia Z, Mark AE, Cooper MA. Developments in glycopeptide antibiotics. *ACS Infect Dis*. 2018;4(5):715–735.
49. Luo Y, Wang C, Peng P, et al. Visible light mediated killing of multidrug-resistant bacteria using photoacids. *J Mater Chem B*. 2013;1(7):997–1001.
50. Li X, Yeh Y-C, Giri K, et al. Control of nanoparticle penetration into biofilms through surface design. *Chem Commun*. 2015;51(2):282–285.
51. Li Q, Wu Y, Lu H, et al. Construction of supramolecular nanoassembly for responsive bacterial elimination and effective bacterial detection. *ACS Appl Mater Interfaces*. 2017;9(11):10180–10189.
52. Weintraub A. Immunology of bacterial polysaccharide antigens. *Carbohydr Res*. 2003;338(23):2539–2547.
53. Hargrove AE, Reyes RN, Riddington I, Anslyn EV, Sessler JL. Boronic acid porphyrin receptor for ginsenoside sensing. *Org Lett*. 2010;12(21):4804–4807.
54. Yan J, Fang H, Wang B. Boronolectins and fluorescent boronolectins: an examination of the detailed chemistry issues important for the design. *Med Res Rev*. 2005;25(5):490–520.
55. Miranda OR, Li X, Garcia-Gonzalez L, et al. Colorimetric bacteria sensing using a supramolecular enzyme-nanoparticle biosensor. *J Am Chem Soc*. 2011;133(25):9650–9653.
56. Zhu C, Yang Q, Liu L, Wang S. Rapid, simple, and high-throughput antimicrobial susceptibility testing and antibiotics screening. *Angew Chem Int Ed Engl*. 2011;50(41):9607–9610.
57. Yuan H, Wang B, Lv F, Liu L, Wang S. Conjugated-polymer-based energy-transfer systems for antimicrobial and anticancer applications. *Adv Mater*. 2014;26(40):6978–6982.
58. Yang W, Fan H, Gao X, et al. The first fluorescent diboronic acid sensor specific for hepatocellular carcinoma cells expressing sialyl Lewis X. *Chem Biol*. 2004;11(4):439–448.
59. Luo J, Xie Z, Lam JWY, et al. Aggregation-induced emission of 1-methyl-1,2,3,4,5-pentaphenylsilole. *Chem Commun*. 2001;(18):1740–1741.
60. Hu R, Zhou F, Zhou T, et al. Specific discrimination of gram-positive bacteria and direct visualization of its infection towards mammalian cells by a DPAN-based AIEgen. *Biomaterials*. 2018;187:47–54.
61. Feng G, Yuan Y, Fang H, et al. A light-up probe with aggregation-induced emission characteristics (AIE) for selective imaging, naked-eye detection and photodynamic killing of Gram-positive bacteria. *Chem Commun*. 2015;51(62):12490–12493.
62. Zhao E, Chen Y, Wang H, et al. Light-enhanced bacterial killing and wash-free imaging based on AIE fluorogen. *ACS Appl Mater Interfaces*. 2015;7(13):7180–7188.
63. Naik VG, Hiremath SD, Das A, et al. Sulfonate-functionalized tetraphenylethylenes for selective detection and wash-free imaging of gram-positive bacteria (*Staphylococcus aureus*). *Mater Chem Front*. 2018;2(11):2091–2097.
64. Kashyap DR, Wang M, Liu LH, Boons GJ, Gupta D, Dziarski R. Peptidoglycan recognition proteins kill bacteria by activating protein-sensing two-component systems. *Nat Med*. 2011;17(6):676–683.
65. Kathiravan A, Sundaravel K, Jaccob M, et al. Pyrene Schiff base: photophysics, aggregation induced emission, and antimicrobial properties. *J Phys Chem B*. 2014;118(47):13573–13581.

66. Zhuk I, Jariwala F, Attygalle AB, Wu Y, Libera MR, Sukhishvili SA. Self-defensive layer-by-layer films with bacteria-triggered antibiotic release. *ACS Nano*. 2014;8(8):7733–7745.
67. Hu D, Deng Y, Jia F, Jin Q, Ji J. Surface charge switchable supramolecular nanocarriers for nitric oxide synergistic photodynamic eradication of biofilms. *ACS Nano*. 2020;14(1):347–359.
68. Yang Y, Reipa V, Liu G, et al. pH-sensitive compounds for selective inhibition of acid-producing bacteria. *ACS Appl Mater Interfaces*. 2018;10(10):8566–8573.
69. Wang J, Gao M, Cui Z-K, et al. One-pot quaternization of dual-responsive poly(vinyl alcohol) with AIEgens for pH-switchable imaging and killing of bacteria. *Mater Chem Front*. 2020;4(9):2635–2645.
70. Lu Y, Huang S, Liu Y, He S, Zhao L, Zeng X. Highly selective and sensitive fluorescent turn-on chemosensor for Al^{3+} based on a novel photoinduced electron transfer approach. *Org Lett*. 2011;13(19):5274–5277.
71. West AP, Brodsky IE, Rahner C, et al. TLR signalling augments macrophage bactericidal activity through mitochondrial ROS. *Nature*. 2011;472(7344):476–480.
72. Hengchang Ma YM, Lei L, Yin W, et al. Light-enhanced bacterial killing and less toxic cell imaging: a multi-cationic AIE matters. *ACS Sustain Chem Eng*. 2018;6:15064–15071.
73. He X, Yang Y, Guo Y, et al. Phage-guided targeting, discriminative imaging, and synergistic killing of bacteria by AIE bioconjugates. *J Am Chem Soc*. 2020;142(8):3959–3969.
74. Gu Y, Zhao Z, Su H, et al. Exploration of biocompatible AIEgens from natural resources. *Chem Sci*. 2018;9(31):6497–6502.
75. Shi X, Yu CYY, Su H, et al. A red-emissive antibody–AIEgen conjugate for turn-on and wash-free imaging of specific cancer cells. *Chem Sci*. 2017;8(10):7014–7024.
76. Yang Z, Chi Z, Mao Z, et al. Recent advances in mechano-responsive luminescence of tetraphenylethylene derivatives with aggregation-induced emission properties. *Mater Chem Front*. 2018;2(5):861–890.
77. Shi H, Liu J, Geng J, Tang BZ, Liu B. Specific detection of integrin αvβ3 by light-up bioprobe with aggregation-induced emission characteristics. *J Am Chem Soc*. 2012;134(23):9569–9572.
78. Gao M, Hu Q, Feng G, et al. A multifunctional probe with aggregation-induced emission characteristics for selective fluorescence imaging and photodynamic killing of bacteria over mammalian cells. *Adv Healthc Mater*. 2015;4(5):659–663.
79. Kumar V, Naik VG, Das A, et al. Synthesis of a series of ethylene glycol modified water-soluble tetrameric TPE-amphiphiles with pyridinium polar heads: towards applications as light-up bioprobes in protein and DNA assay, and wash-free imaging of bacteria. *Tetrahedron*. 2019;75(27):3722–3732.
80. Feng G, Zhang C-J, Lu X, Liu B. Zinc(II)-tetradentate-coordinated probe with aggregation-induced emission characteristics for selective imaging and photoinactivation of bacteria. *ACS Omega*. 2017;2(2):546–553.
81. Li NN, Li JZ, Liu P, et al. An antimicrobial peptide with an aggregation-induced emission (AIE) luminogen for studying bacterial membrane interactions and antibacterial actions. *Chem Commun*. 2017;53(23):3315–3318.
82. Shi J, Wang M, Sun Z, et al. Aggregation-induced emission-based ionic liquids for bacterial killing, imaging, cell labeling, and bacterial detection in blood cells. *Acta Biomater*. 2019;97:247–259.
83. Kang M, Kwok RTK, Wang J, et al. A multifunctional luminogen with aggregation-induced emission characteristics for selective imaging and photodynamic killing of both cancer cells and gram-positive bacteria. *J Mater Chem B*. 2018;6(23):3894–3903.
84. Panigrahi A, Are VN, Jain S, Nayak D, Giri S, Sarma TK. Cationic organic nano-aggregates as AIE luminogens for wash-free imaging of bacteria and broad-spectrum antimicrobial application. *ACS Appl Mater Interfaces*. 2020;12(5):5389–5402.

CHAPTER FOUR

Aggregation-induced emission materials for cell membrane imaging

Dipratn G. Khandare*

Department of Chemistry, Modern College of Arts, Science and Commerce Ganeshkhind, Pune, India
*Corresponding author: e-mail address: dipratnkhandare7885@gmail.com

Contents

1. Introduction	82
2. Difficulties in using traditional fluorescent sensors	83
3. Aggregation-induced emission (AIE) phenomenon	84
4. Mechanism of aggregation induced emission (AIE) effect	85
5. Examples of cell membrane imaging	86
5.1 Detection of Cu^{2+} in live cells and cell membrane imaging	88
5.2 Fluorescence cell membrane imaging	92
6. Conclusion	94
7. Future scope	94
References	95

Abstract

The living cells are consist of protective layer, which is called cell membrane. The cell membrane is associated with numerous biological functions and they are intensely depended on fundamental physicochemical properties of cell membrane. The cell metabolism and functions can be studied by investigating morphology of cell membrane. Cell membrane disruption causes depolarization due to which cell content leakage and cell death can occur. The fluorescence technique has been extensively used to monitor intracellular structure and for cell membrane imaging. Tang's group reported a group of fluorophore/luminophore which emit more light in the aggregated form than in solution. Aggregate formation played constructive role in this kind of luminogens in the light emitting phenomenon. The aggregated form of reported series of silole molecules/fluorophores were found to be more emissive rather than in solution. These luminophores showed exactly opposite emission behavior to ACQ and this phenomenon coined as "Aggregation-Induced Emission" (AIE). In this chapter we have given a brief review of the AIE materials for cell membrane imaging.

Graphical abstract

General graphical representation showing an Aggregation-Induced Emission after interacting with analyte/target.

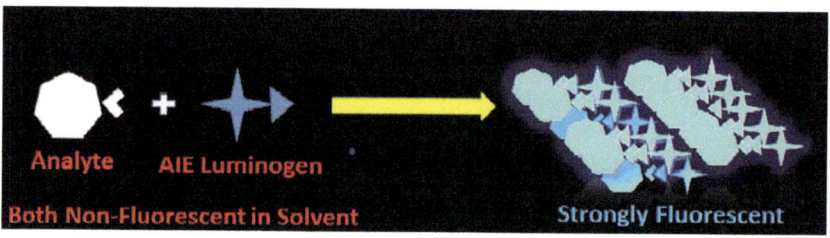

1. Introduction

The living cells are consist of protective layer, which is called cell membrane. The cell membrane is associated with numerous biological functions and they are intensely depended on fundamental physicochemical properties of cell membrane. The cell membrane consists of proteins, lipids (cholesterol and phospholipids) and glycoproteins and glycolipids as a carbohydrate residues. The membrane constituting molecules define the membrane structure and regulate the binding and transport of ionic and molecular species. Cell membrane particularly is involved in various cellular functions like signal transduction, nutrient transport, exocytosis and endocytosis etc.[1–3] It does not allow toxic material to enter inside the cell. Because of not having homogeneous distribution of lipids and proteins plasma membrane of eukaryotic cells, gives rise to biophysical and biochemical different regions. Lipid rafts are the highly ordered region and have been assumed to exist in the cell membrane for protein diffusion and distribution.[4–7] The cell metabolism and functions can be studied by investigating morphology of cell membrane.[8–10] Cell membrane disruption causes depolarization due to which cell content leakage and cell death can occur. Therefore, it is extremely important to monitor real-time morphological change in cell membrane in different conditions in drug development and biomedical research.

The development of new technique like fluorescence imaging is used as powerful and reliable technique which helped to develop and synthesize fluorescent sensors to understand and monitor biological targets and subcellular localization with its sensitivity, selectivity and rapidness.[11–15] The fluorescence technique has been extensively used to monitor intracellular structure and for cell membrane imaging.[16,17] A derivative of Prodan, Laurdan (6-lauryl-2-dimethylamino-naphthalene) is a most popular polarity sensitive dye used for exploration of membrane organization reported by Weber and Farris in 1979.[18] With no adverse effect on membrane Laurdan has been used to stain whole organisms, artificial cells as well as fixed and live cells.[19,20] Laurdan, in spite of its vast use it has significant disadvantages because irradiation of UV can cause phototoxic effect on live cells.

2. Difficulties in using traditional fluorescent sensors

The real world applications of traditional fluorophores limited by the phenomenon called aggregation caused quenching (ACQ) effect.[21,22] Forster and Kasper, in 1953 reported that the fluorescence is weakened in case of polyaromatic compounds, with increase in concentration of solution.[22,23] Excimers and exciplex formation is responsible factor for the quenching of fluorescence, which is called as aggregation caused quenching (ACQ). Later on this phenomenon became general for many aromatic systems.[23] Fluorescence quenching caused by increase in concentration of fluorophore by the formation of excimers and exciplexes. Excimers and exciplexes formation is then result of collisional interaction between aromatic molecules in ground and excited state.[22a] Traditional fluorophore show weak fluorescence in dilute solutions however use of dilute solution lead to the very weak fluorescence which cause poor sensitivity in the fluorescence system, especially for bioassay, when dilute solution is employed for the detection of trace amount of biomolecules.[24] Use of high concentration of fluorophore, cannot increase the sensitivity, because high concentration causes fluorescence quenching effect (ACQ). Whereas, use of dilute solution also causes ACQ, use of dilute solution of fluorophore in bioassay, causes accumulation of small fluorophore molecules on surface of biomolecules. Fluorophore accumulation increases the local concentration of fluorophore which results into concentration quenching. Because "formation of aggregates" is the major reason for the quenching.[25,26]

Traditional fluorophores in solid state, fluorophore molecules are in immediate vicinity to each other. Because of which, aromatic rings of neighboring molecules experience π-π stacking interaction, which causes the formation of excimers. When these excimers are in excited state, they decay via non-radiative pathway, called aggregation caused quenching (ACQ) in the concentrated solutions or in solid state.[27]

Fluorophore/luminophore behavior is studied generally in solution state and in real world they are used as material, like in fluorescent diagnostic kits,[24,25] and thin films in the fabrications of organic-light emitting diodes (OLED's).[28,29] Many leading luminogens/fluorophores due to ACQ effect cannot be used in the solid state as it reduced the fluorescence, ultimately which limits their real world application. Many researcher, to overcome the problem of ACQ, have developed many approaches. To avoid the aggregate formation researchers used the strategies like covalent attachment of bulky chains, spiro kinks and dendritic wedges on the aromatic rings.[30] But these strategies proved to be the destructive, because attachment of non-conjugated groups and sterically bulky groups to the aromatic rings, can twist the conformation and thereby destructively affecting the π-conjugation. Researchers despite of using such approaches ended up with the limited success, because aggregate formation is the intrinsic property. Many of the approaches blocked aggregate formation temporarily and partially.[30,31] Therefore, there was a need to develop fluorophore/luminophore which can emit light in the concentrated solution or in solid state.

3. Aggregation-induced emission (AIE) phenomenon

Researchers started focusing on to develop such luminogens or system which can emit more light in aggregated state rather than quenching. In 2001, Tang's group reported a group of fluorophore/luminophore (Fig. 1) which emit more light in the aggregated form than in solution. Aggregate

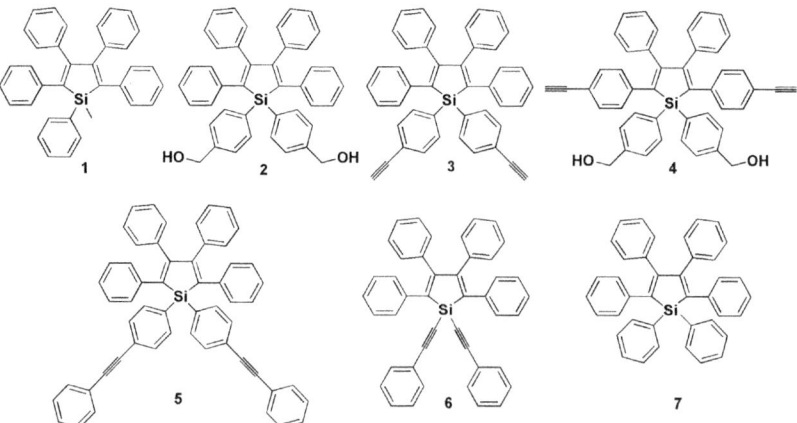

Fig. 1 Series of silole luminogens (**1–7**), which show AIE phenomenon.

formation played constructive role in this kind of luminogens in the light emitting phenomenon. The aggregated form of reported series of silole derivatives were found to be more emissive rather than in solution.[32] These luminophores showed exactly opposite emission behavior to ACQ and this phenomenon coined as "Aggregation-Induced Emission" (AIE).[32,33] This phenomenon has its real world applications, by taking the advantage of fluorophoric aggregate formation one can use dye solution at higher concentration for bioassay to develop "light-up" or "turn-on" type sensors.

4. Mechanism of aggregation induced emission (AIE) effect

Tang's group in 2001, demonstrated the mechanism of AIE by taking an example of Hexaphenylsilole (HPS, **9**) molecule. Restriction in intramolecular rotation is the major factor responsible for aggregation induced emission (AIE) effect.[32b] When HPS molecules are dissolved in the solvents like THF or acetonitrile, they are completely non-emissive in the solution. With increase in the concentration of water in HPS solution, they start to aggregate which leads to the strong emission from the system. Geometrically non-planar hexaphenylsilole (HPS) molecule possesses six phenyl rings, which gives propeller shape to the molecule. In the solution or in dissolved form, six aromatic rings in the HPS rotate through the single bond axes and along the side of silole stator. When the HPS molecules are in excited state in the solution, the rotation through single bond acts as relaxation channel and the molecule follow the non-radiative decay. Intramolecular rotation in AIE molecules converts the photonic energy to heat which causes non-radiative relaxation of excited state.

HPS molecules upon addition of poor solvent like water start aggregating which leads to restriction in intramolecular rotation in the solution. Due to the aggregation of HPS molecules, relaxation of excited state follows radiative pathway and the system shows strong emission (Fig. 2).

Apart from Hexaphenylsilole and its derivatives, there are various other molecules (numbers) which show aggregation-induced emission (AIE) property, they are completely non-emissive in good solvent and they exhibit strong emission in aggregated form (Fig. 2).[34,35] The structures mentioned in Fig. 3, all have carbons and hydrogens, neither they have heteroatoms nor planar groups present in them. Therefore they cannot form H-aggregates and J-aggregates and TICT effect in their luminescence process.

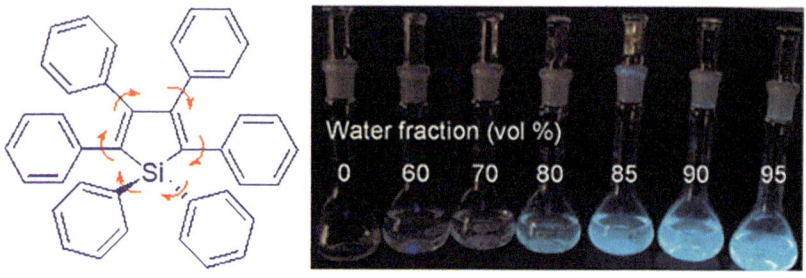

Hexaphenylsilole (HPS, 7)

Fig. 2 Structure of Hexaphenylsilole (HPS, **7**) (left) and solution of HPS (right) in different vol% of acetonitrile-water mixture.

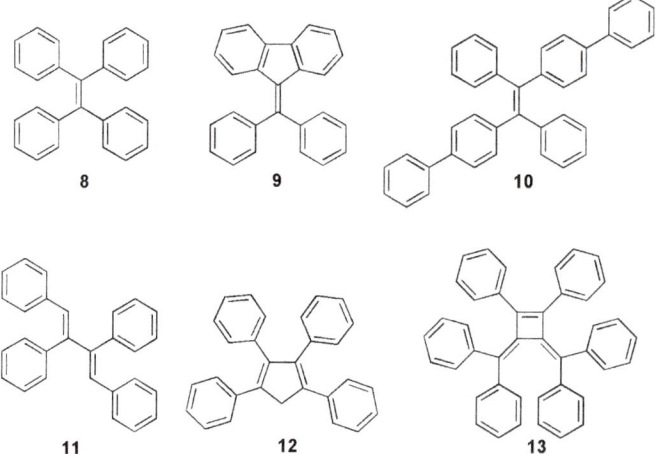

Fig. 3 Chemical structures of some luminogens (**8–13**) which show AIE property.

Because of unique AIE property of these luminogens researchers started using them in various fields like for the electroluminescence devices, chemical sensing biomedical applications, explosive sensing, pH sensing, molecular sensing and cell imaging.[36] Among which cell membrane imaging is one of the important application of AIE luminogens. Following are the few examples for cell membrane imaging using aggregation-induced emission (AIE) based luminogens.

5. Examples of cell membrane imaging

Fuyou Li and his co-workers in 2013 reported a aggregation-induced emission based bioprobe 1,8-naphthalimide derivative (FD-9), a new organic dye for cell membrane tracking.[37] It specifically adheres and tracks cell membrane. For the probe FD-9, cell bioimaging is studied with HepG-2 cells where it showed negligible background fluorescence. But FD-9 after incubating with living HepG-2 cells intense fluorescence was observed in the membrane region. Intracellular fluorescence emission from 400 to 600 nm is generated from FD-9 after incubating it with HepG-2 cells. Further the cell membrane staining was confirmed by intramolecular colocalization experiment with commercially available cell membrane imaging dye DiI. The fluorescent images (Fig. 4) obtained from colocalization experiment were overlapped, FD-9 (green) and colocalization dye (red) and 3D visualization image.

In 2017, Dan Ding and his co-workers, reported cell membered anchored AIEgen based probe named TPE-Py-EEGTIGYG (Fig. 5) for detection Cu^{2+}, the probe was synthesized by using solid phase synthesis

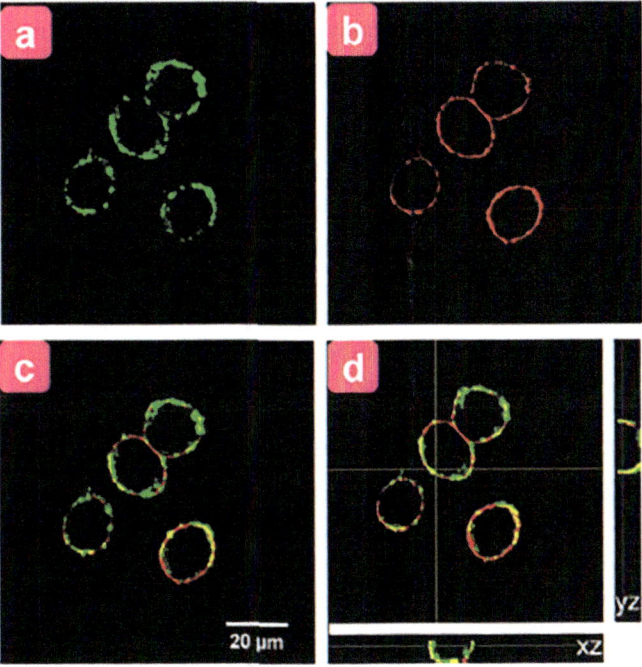

Fig. 4 The probe FD-9 (5 mM) incubated with living HepG-2 cells for 30 min and 5 mM of DiI for 15 min in DMSO–RPMI-1640 (1:49, v/v) at 37 °C. (A) Confocal fluorescence image by making use of FD-9 (green, $\lambda_{ex}=405$ nm, $\lambda_{em}=480.50$ nm). (B) Confocal fluorescence image by making use of DiI (red, $\lambda_{ex}=543$ nm, $\lambda_{em}=600.50$ nm). (C) Overlap of (A) and (B). (D) 3D fluorescence images of (C).

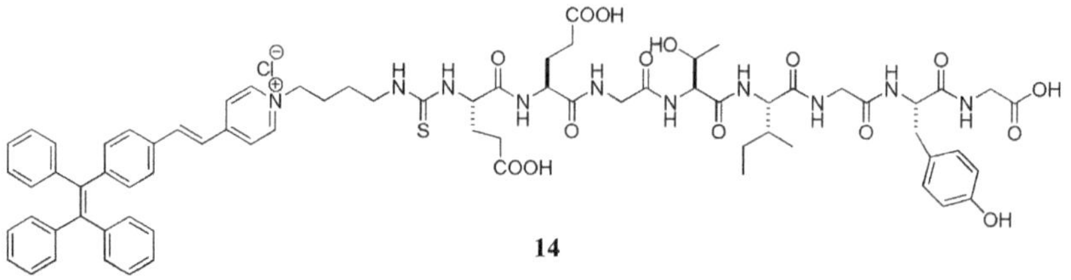

Fig. 5 Structure of AIEgen fluorescent probe TPE-Py-EEGTIGYG.

of peptide and later addition reaction of the isothiocyanate group of TPEPy-NCS and the N-terminal amino group of NH_2-EEGTIGYG.[38] The advantage of this fluorescent probe is, it is highly selective for the detection of divalent Cu^{2+} in live organisms as well as in aqueous environment. The is consisting of two parts a) AIEgenic TPE-Py to impart fluorescence to the probe and b) short peptide linkage EEGTIGYG (E: glutamic acid; G: glycine; T: Threonine; I: isoleucine; Y: Tyrosine). Presence of quaternary ammonium salt and hydrophilic peptide linkage in the probe increases the water solubility and specificity in binding to the cell membrane.

TPE-Py-EEGTIGYG is weakly fluorescent in aqueous medium as it exists as molecular species. On the other hand which upon anchoring on cell membrane shows strong fluorescence. Many diseases are associated with the movement of Cu^{2+} in and out of the cells. Due to which it is extremely important to monitor movement of Cu^{2+}. This probe can selectively detects the Cu^{2+} by "turn-off" type fluorescence signal. Which could give valuable information to clinicians and scientists about Cu^{2+} related diseases.

Further the cytotoxicity of TPE-Py-NCS and TPE-Py-EEGTIGYG was evaluated against HeLa cells by MTT assay. The cell viability after incubation for 48 h was found to be around 80–90%.

5.1 Detection of Cu^{2+} in live cells and cell membrane imaging

In selectivity study, 25 μM of concentration was used for detection Cu^{2+} using probe TPE-Py-EEGTIGYG in aqueous media. This study revealed that with increase in concentration of Cu^{2+}, fluorescence intensity was found to be decreased for nanoaggregates of TPE-Py-EEGTIGYG. TPE-Py-EEGTIGYG lost about 60% of fluorescence intensity upon addition of 8 μM of Cu^{2+}. Upon further increasing the concentration of Cu^{2+} to 1.5 mM, fluorescence intensity decreased by about 85%. This study showed that TPE-Py-EEGTIGYG can selectively be used as nanoprobe for detection of Cu^{2+} in aqueous medium.

Further the selectivity of the probe was evaluated by incubating the probe nanoaggregates (25 μM) with different competing metal ions like K^+, Zn^{2+}, Fe^{2+}, Ca^{2+}, Co^{2+}, Pb^{2+}, Mn^{2+}, Ba^{2+}, Mg^{2+}, Na^+, Fe^{3+}, Ni^{2+}, Ag^+, Hg^{2+}, and Cu^{2+} with constant concentration of each metal ions (16 μM), in PBS buffer. From this study it was revealed that only Cu^{2+} can selectively quench the fluorescence intensity of nanoaggregates of TPE-Py-EEGTIGYG.

In the next part of study, Cu^{2+} detection and cell membrane imaging in live cell by TPE-Py-EEGTIGYG was carried out with confocal laser scanning microscopy (CLSM). 5 μM of TPE-Py-EEGTIGYG was used in the cellular experiment because at lower concentration the probe exist as weakly fluorescent and molecular species in aqueous media. It was assumed that the interaction between cell membrane and TPE-Py-EEGTIGYG could restrict the intramolecular rotation of phenyl rings of the AIEgenic probe which would result into the fluorescence "turn-on" on the cell membrane. The above hypothesis was confirmed by incubating HeLa cancer cells with 5 μM concentration of TPE-Py-EEGTIGYG for 2 h and imaging live HeLa cancer cells.

A strong yellow fluorescence was observed from cell membrane anchored with TPE-Py-EEGTIGYG (Fig. 6A). The HeLa cancer cells were

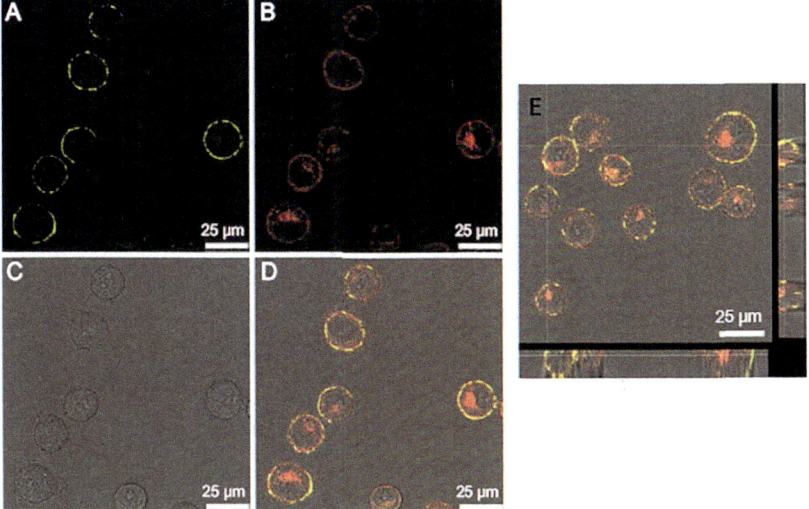

Fig. 6 CLSM images (A) HeLa cells anchored with TPE-Py-EEGTIGYG (5 μM) and (B) HeLa cells co-stained with DiI, (C) transmission image, (D) overlay images of A, B and C. (E) Sectional 3D CLSM image of TPE-Py-EEGTIGYG (5 μM) and co-stained with DiI HeLa cells.

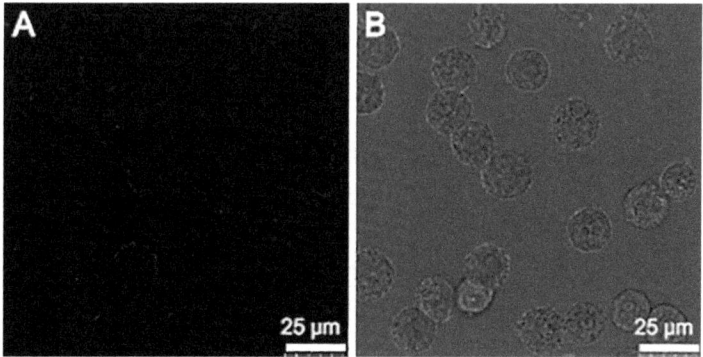

Fig. 7 CLSM image of (A) HeLa cells stained with TPE-Py-EEGTIGYG (5 μM) after addition of Cu^{2+} (1.5 mM). (B) Corresponding transmission/fluorescence overlay image of (A).

co-stained with commercially available cell membrane-labelling agent DiI (Fig. 6B). Fig. 6D and E (3D section of CLSM image of TPE-Py-EEGTIGYG) showed the good overlap of fluorescence signal from the cellular membrane of HeLa cells anchored with TPE-PyEEGTIGYG and DiI respectively. The Cu^{2+} (1.5 mM) was added to the already stained cell membrane with TPE-Py-EEGTIGYG. As shown in Fig. 7, the fluorescence intensity was greatly quenched upon addition of Cu^{2+}. Which revealed that, the TPE-Py-EEGTIGYG can be used as a tool for detecting Cu^{2+} in live cells.

In 2018, B. Z. Tang and his co-workers developed a photostable lipophilic AIEgen with near-infrared emission, highly biocompatible and highly photostable called cyanostilbene (AS2CP-TPA) for cell membrane imaging and monitoring membrane morphology under different circumstances (Fig. 8).[39] The probe (AS2CP-TPA) comprised of a hydrophobic triphenylamino group and a hydrophilic pyridinium salt. The probe (AS2CP-TPA) shows acceptor-donor interaction as well as aggregation-induced emission property. Due to which it shows weak emission at 460 nm wavelength in the aqueous solution but it emits strong light near infrared when attached with the plasma membrane of HeLa cells. The probe (AS2CP-TPA) possesses high specificity and high photostability toward the plasma membrane. Because of its photostability, it can be used in the treatments of Hg^{2+} and trypsin for monitoring morphological change of cell membrane.

It was observed from the MTT assay that the probe AS2CP-TPA possesses low toxicity because 80% of HeLa cells showed viability after incubating with AS2CP-TPA. Which proved that it is highly biocompatible and less cytotoxic and therefore suitable for cell imaging. To assess the selectivity

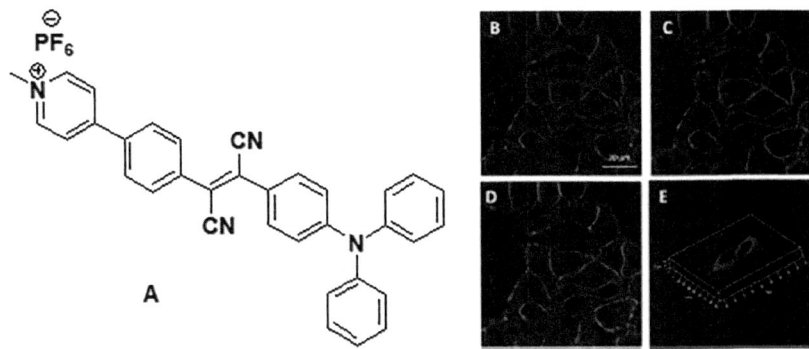

Fig. 8 (A) Chemical structure of AS2CP-TPA. 2B-E are the (B–D) HeLa cells confocal images co-stained with probe AS2CP-TPA (3.5 µM) and CellMask Green Plasma Membrane (5 ng/mL) for 5 min. Conditions: (B) λ_{ex}: 526 nm and λ_{em}: 600–700 nm for AS2CP-TPA; (C) λ_{ex}: 488 nm and λ_{em}: 500–540 nm for CellMask Green Plasma Membrane. (D) The overlay images of (B) and (C). (E) 3D confocal images of HeLa cell stained with probe AS2CP-TPA obtained by scanning different layers. Condition: λ_{ex}: 488 nm and λ_{em}: 600–750 nm.

of AS2CP-TPA toward the plasma membrane of HeLa cells, a commercially available biomarker CellMask Green Plasma Membrane was used to co-stain the HeLa cells along with AS2CP-TPA (Fig. 8B–D). Which disclosed that AS2CP-TPA is specifically stained the plasma membrane. The specificity of AS2CP-TPA was further verified dye stained 3D structure of HeLa cells was obtained by scanning different layers with confocal microscope (Fig. 8E). It was concluded from the HeLa cell membrane scanned images that cell membrane was well visualized and cytoplasmic components of the HeLa cell was entirely non-fluorescent. This result demonstrated that the AS2CP-TPA is plasma membrane specific as like the commercially available CellMask Green Plasma Membrane biomarker.

Further by continuous scanning the dye stained HeLa cells the photostability of the AS2CP-TPA was investigated and compared with CellMask Deep Red Plasma Membrane. Which revealed that after 50 scans about 20% of fluorescent signal of AS2CP-TPA was lost in the same excitation power. Whereas the fluorescence intensity of CellMask Deep Red Plasma Membrane was completely quenched. This showed that the probe AS2CP-TPA possesses high resistance for photo-oxidation and therefore it can be used for the monitoring morphology of plasma membrane.

In conclusion highly photostable, biocompatible near-infrared AIEgen AS2CP-TPA was synthesized. The photostability of the AIEgen is useful for the long term monitoring of changes in cell morphology in different conditions for example detachment of adherent cells by trypsin or Hg^{2+} treatment with live cells.

A cell membrane based probe TPE conjugated with a coumarin through α, β-unsaturation was reported by Yibin Zhang and his co-workers in 2020.[40] The aggregation-induced emission property of the probe was investigated in tetrahydrofuran (THF) solvent. Further, sensitivity of the probe was investigated. From this study it was found that, probe showed slight change in the absorbance and fluorescence with increase in pH value from 4.0 to 9.2. Probe did not show any significant change in the fluorescence intensity when treated with competing anions such as SO_4^{2-}, NO_3^-, HCO_3^-, PO_4^{3-}, SO_3^{2-}, and CO_3^{2-}, and reactive nitrogen, oxygen, and sulfur species, such as cysteine (Cys), glutathione (GSH), homocysteine (Hcy), HSO_3^-, $S_2O_3^{2-}$, H_2S, H_2O_2, O_2^-, $ONOO^-$, and ClO^- and cations like Zn^{2+}, Ag^+, Mg^{2+}, Co^{2+}, Mg^{2+}, Al^{3+}, Pb^{2+}, Fe^{2+}, K^+, Hg^{2+}, Cr^{3+} and Fe^{3+}. This result suggested that the probe is insensitive toward these interfering anions and cations. The photostability of probe was tested by conducting photostability experiment and compared with CellBrite cell membrane NIR dye. After exciting probe at 488 nm and CellBrite cell membrane NIR dye at 630 nm continuously for 40 min. It was found that the probe fluorescence intensity was decreased by just 2.5%. This means probe possesses a very good photostability.

The cytotoxicity of the probe was investigated by using MTT call viability assay. From this study it was found that the cell viability was found to be 86% with 50 μM concentration of the probe after 24 h. Which suggested that the probe has low toxicity at higher concentration and can be utilized for staining live cell.

5.2 Fluorescence cell membrane imaging

Further to demonstrate, the cell membrane imaging and the probe is specific to cell membrane, colocalization experiment was carried out. In which HeLa cells were incubated for 15 min with the TPE conjugated with a coumarin through α, β-unsaturation and CellBrite cell membrane commercially available NIR dye (Fig. 9). It was observed from this experiment, bright fluorescence was shown at 405 nm excitation around the cell membrane and at 488 nm strong fluorescence was observed around the cell membrane in the pink channel. The colocalization experiment was conducted

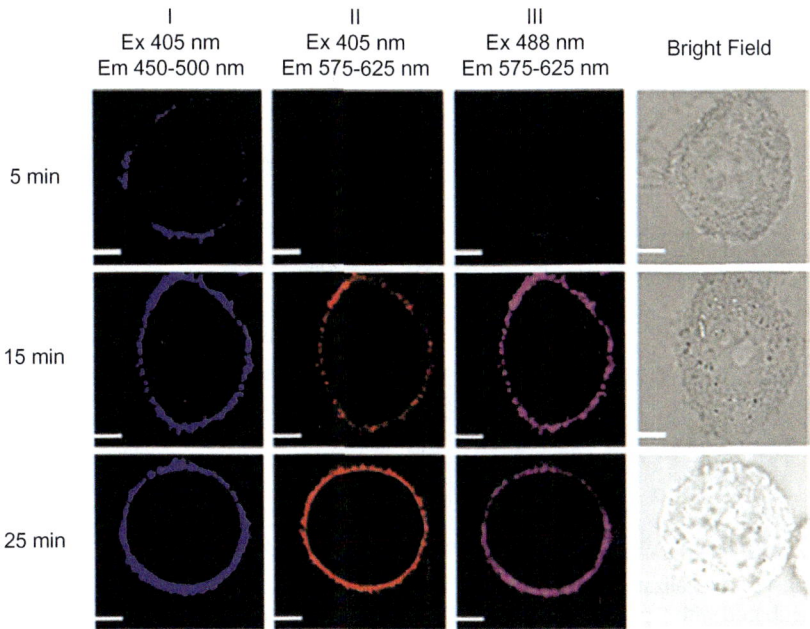

Fig. 9 Fluorescence images of HeLa cell after addition of probe (10 μM) excitations at 405 and 488 nm after incubating with HeLa cells under different incubation time. A confocal fluorescence microscope (Olympus IX 81) was used to capture the fluorescence images.

further to confirm the fluorescence is from the probe. The CellBrite cell membrane NIR dye was used to label specifically cell membrane and it was merged with the probe, fluorescence in the blue channel under 630 nm excitation. This demonstrated that the blue fluorescence is of probe which colocalizes with the CellBrite Cell membrane NIR dye. This confirmed that the probe can specifically be used for cell membrane staining.

The probe was further employed for staining breast cancer MCF-7 cells to determine the probe AS2CP-TPA can stain other cells also. In this colocalization experiment probe and breast cancer MCF-7 cells was incubated for 15 min along with commercial membrane-specific probe (CellBrite cytoplasmic NIR dye) under excitation at 405 and 488 nm for probe A and commercial dye at 630 nm (Fig. 10). This study revealed that the probe is target specific and stains only cell membrane of breast cancer MCF-7 cells, which gave 0.93 Pearson correlation coefficient with commercial membrane-specific probe.

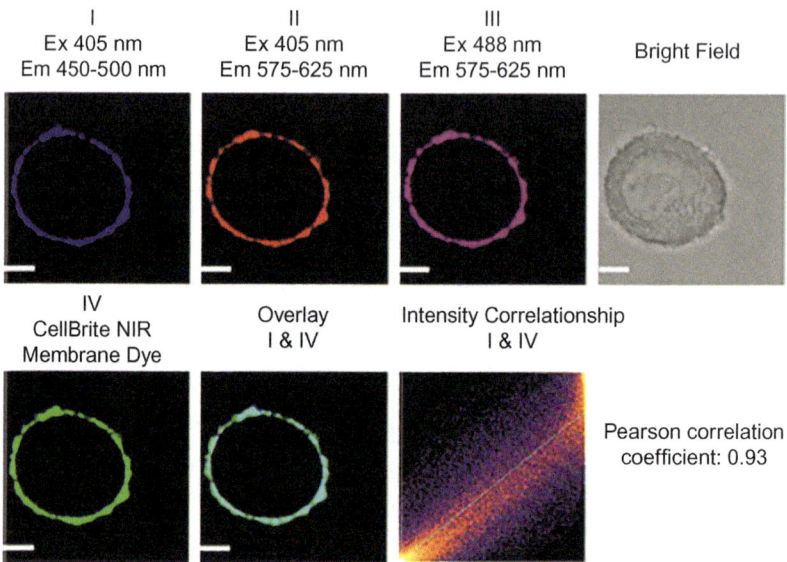

Fig. 10 Colocalization experiment fluorescence images of a MCF-7, breast cancer cell incubated with 5 μM CellBrite cytoplasmic NIR dye and 10 μM probe A under excitation of 488 nm and 405 for the probe, and 630 nm for CellBrite cytoplasmic NIR dye. 0.93 was the Pearson correlation coefficient of channel I with channel IV.

6. Conclusion

In conclusion, a variety of chemosensors/chemodosimeters have been synthesized by making use of AIE-active framework. By incorporating different functional groups in AIE-active framework which can provide different chemodosimeters/chemosensors and AIE-active materials, which interact with the analytes or target covalently or non-covalently forming self-assembled aggregates. In this chapter we have briefly given a short review of aggregation-induced emission materials for cell membrane imaging.

7. Future scope

Though there are many reports available in literature on AIE-active compounds but still there are some areas which are unexplored. The current development depicts the great potential of aggregation-induced emission (AIE) based materials due to its unique property, they can be utilized for cell membrane imaging, also can be further utilized for the biological sensing, biomedical applications and for many more applications like pH sensing, in solar cells, electrochromic devices, cation/anion sensing, explosive sensing, in OLED's etc.

References

1. Schlessinger J, Axelrod D, Koppel D, Webb W, Elson E. Lateral transport of a lipid probe and labeled proteins on a cell membrane. *Science*. 1977;195:307–309.
2. Grimmer S, Van Deurs B, Sandvig K. Membrane ruffling and macropinocytosis in A431 cells require cholesterol. *J Cell Sci*. 2002;115:2953–2962.
3. Rejman J, Oberle V, Zuhorn IS, Hoekstra D. Size-dependent internalization of particles via the pathways of clathrin- and caveolae-mediated endocytosis. *Biochem J*. 2004; 377:159–169.
4. Simons K, Ikonen E. Functional rafts in cell membranes. *Nature*. 1997;387:569–572.
5. Pike LJ. Rafts defined: a report on the keystone symposium on lipid rafts and cell function. *J Lipid Res*. 2006;47:1597–1598.
6. Simons K, Gerl MJ. Revitalizing membrane rafts: new tools and insights. *Nat Rev Mol Cell Biol*. 2010;11:688–699.
7. Lingwood D, Simons K. Lipid rafts as a membrane-organizing principle. *Science*. 2010;327:46–50.
8. Zwaal RF, Schroit AJ. Pathophysiologic implications of membrane phospholipid asymmetry in blood cells. *Blood*. 1997;89:1121–1132.
9. Fadeel B, Xue D. The ins and outs of phospholipid asymmetry in the plasma membrane: roles in health and disease. *Crit Rev Biochem Mol Biol*. 2009;44:264–277.
10. Gutknecht J. Inorganic mercury (Hg2+) transport through lipid bilayer membranes. *J Membr Biol*. 1981;61:61–66.
11. Zhu H, Fan JL, Du JJ, Peng XJ. Fluorescent probes for sensing and imaging within specific cellular organelles. *Acc Chem Res*. 2016;49:2115–2126.
12. Gao P, Pan W, Li N, Tang B. Fluorescent probes for organelle-targeted bioactive species imaging. *Chem Sci*. 2019;10:6035–6071.
13. Xu W, Zeng ZB, Jiang JH, Chang YT, Yuan L. Discerning the chemistry in individual organelles with small-molecule fluorescent probes. *Angew Chem, Int Ed*. 2016;55: 13658–13699.
14. de Silva AP, Gunaratne HQN, Gunnlaugsson T, et al. Signaling recognition events with fluorescent sensors and switches. *Chem Rev*. 1997;97:1515–-1566.
15. Lim X. The nanoscale rainbow. *Nature*. 2016;531:26–28.
16. Owen DM, Neil MAA, French PMW, Magee AI. Optical techniques for imaging membrane lipid microdomains in living cells. *Semin Cell Dev Biol*. 2007;18:591–598.
17. Lagerholm BC, Weinreb GE, Jacobson K, Thompson NL. Detecting microdomains in intact cell membranes. *Annu Rev Phys Chem*. 2005;56:309–336.
18. Weber G, Farris FJ. Synthesis and spectral properties of a hydrophobic fluorescent probe: 6-propionyl-2-(dimethylamino)naphthalene. *Biochemistry*. 1979;18:3075–3078.
19. Owen DM, Magenau A, Majumdar A, Gaus K. Imaging membrane lipid order in whole, living vertebrate organisms. *Biophys J*. 2010;99:L7–L9.
20. Owen DM, Rentero C, Magenau A, Abu-Siniyeh A, Gaus K. Quantitative imaging of membrane lipid order in cells and organisms. *Nat Protoc*. 2012;7:24–35.
21. (a) Duke RM, Veale EB, Pfeffer FM, Kruger PE, Gunnlaugsson T. Colorimetric and fluorescent anion sensors: an overview of recent developments in the use of 1,8-naphthalimide-based chemosensors. *Chem Soc Rev*. 2010;39:3936–3953. (b) Culzoni MJ, Munoz de la Pena A, Machuca A, Goicoechea HC, Babiano R. Rhodamine and

BODIPY chemodosimeters and chemosensors for the detection of Hg^{2+}, based on fluorescence enhancement effects. *Anal Methods.* 2013;5:30–49. (c) Kim HN, Ren WX, Kim JS, Yoon J. Fluorescent and colorimetric sensors for detection of lead, cadmium, and mercury ions. *Chem Soc Rev.* 2012;41:3210–3244. (d) Jun ME, Roy B, Ahn KH. "Turn-on" fluorescent sensing with "reactive" probes. *Chem Commun.* 2011;47:7583–7601. (e) Long F, Zhu A, Shi H, Wang H, Liu J. Rapid on-site/*in-situ* detection of heavy metal ions in environmental water using a structure-switching DNA optical biosensor. *Sci Rep.* 2013;3:2308. (f) Li M, Gou H, Al-Ogaidi I, Wu N. Nanostructured sensors for detection of heavy metals: a review. *ACS Sustainable Chem Eng.* 2013;1:713–723. (g) Chan J, Dodanil SC, Chang CJ. Reaction-based small-molecule fluorescent probes for chemoselective bioimaging. *Nat Chem.* 2012;4:973–984. (h) Chen X, Zhou G, Peng X, Yoon J. Biosensors and chemosensors based on the optical responses of polydiacetylenes. *Chem Soc Rev.* 2012; 41:4610–4630.
22. (a) Brirks JB. *Photophysics of Aromatic Molecules.* vol 74. London: Wiley; 1970:1294–1295. (b) Tong H, Hong Y, Dong Y, et al. Fluorescent "light-up" bioprobes based on tetraphenylethylene derivatives with aggregation-induced emission characteristics. *Chem Commun.* 2006;3705–3707.
23. Forster T, Kasper K. *Phys Chem, Munich.* 1954;1:275.
24. (a) Geddes CD, Lakopwicz JR. *Topics in Fluorescence Spectroscopy, Advanced Concepts in Fluorescence Sensing.* Norwell: Springer; 2005. (b) Thompson RB, ed. *Bioscience, Physical Sciences, Fluorescence Sensors and Biosensors.* vol. 1. Boca Raton: CRC; 2006:1–416. (c) Tan WH, Wang KM, Drake TJ. Molecular beacons. *Curr Opin Chem Biol.* 2004;8:547–553. (d) Sapsford KE, Berti L, Medintz IL. Materials for fluorescence resonance energy transfer analysis: beyond traditional donor–acceptor combinations. *Angew Chem, Int Ed.* 2006;45:4562–4589. (e) Borisov SM, Wolfbeis OS. Optical biosensors. *Chem Rev.* 2008;108:423–461.
25. (a) Domaille DW, Que EL, Chang CJ. Synthetic fluorescent sensors for studying the cell biology of metals. *Nat Chem Biol.* 2008;4:168–175. (b) Lim MH, Lippard SJ. Metal-based turn-on fluorescent probes for sensing nitric oxide. *Acc Chem Res.* 2007;40:41–51. (c) Giepmans BNG, Adams SR, Ellisman MH, Tsien RY. The fluorescent toolbox for assessing protein location and function. *Science.* 2006;312:217–224. (d) Jares-Erijman EA, Jovin TM. FRET imaging. *Nat Biotechnol.* 2003;21:1387–1395.
26. (a) Slavik J. *Fluorescence Microscopy and Fluorescent Probes.* New York: Plenum; 1996. (b) Valeur B. *Molecular Fluorescence: Principle and Applications.* Weinheim: Wiley-VCH; 2001.
27. (a) Thomas III SW, Joly GD, Swager TM. Chemical sensors based on amplifying fluorescent conjugated polymers. *Chem Rev.* 2007;107:1339–1386. (b) Belletete M, Bouchard J, Leclerc, Durocher G. Photophysics and solvent-induced aggregation of 2,7-carbazole-based conjugated polymers. *Macromolecules.* 2005;38:880–887. (c) Menon A, Galvin M, Walz KA, Rothberg L. Structural basis for the spectroscopy and photophysics of solution-aggregated conjugated polymers. *Synth Met.* 2004;141:197–202. (d) Chen C-T. Evolution of red organic light-emitting diodes: materials and devices. *Chem Mater.* 2004;16:4389–4400. (e) Grell M, Bradley DDC, Ungar G, Hill J, Whitehead KS. Interplay of physical structure and photophysics for a liquid crystalline polyfluorene. *Macromolecules.* 1999;32:5810–5817. (f) Jakubiak R, Collison CJ, Wan WC, Rothberg L, Hsieh BR. Aggregation quenching of luminescence in electroluminescent conjugated polymers. *J Phys Chem A.* 1999;103:2394–2398. (g) Grell M, Bradley DDC, Long X, et al. Chain geometry, solution aggregation and enhanced dichroism in the liquidcrystalline conjugated polymer poly(9,9-dioctylfluorene). *Acta Polym.* 1998;49:439–444. (h) Lemmer U, Heun S, Mahrt RF, et al. Aggregate fluorescence in conjugated polymers. *Chem Phys Lett.* 1995;240:373–378.

28. (a) Grimsdale AC, Mullen K. Poly(2,7-carbazole)s and related polymers. *Adv Polym Sci.* 2008;212:99–124. (b) Wong WWH, Holmes AB. Polydibenzosiloles. *Adv Polym Sci.* 2008;212:85–98. (c) Sanchez JC, Trogler WC. Hydrosilylation of diynes as a route to functional polymers delocalized through silicon. *Macromol Chem Phys.* 2008;209:1527–1540. (d) Zhang Y, Liu B, Cao Y. Synthesis and characterization of a water-soluble carboxylated polyfluorene and its fluorescence quenching by cationic quenchers and proteins. *Chem Asian J.* 2008;3:739–745. (e) Chen J, Cao Y. Silole-containing polymers: chemistry and optoelectronic properties. *Macromol Rapid Commun.* 2007;28:1714–1742. (f) Yam VWW, Wong KMC. Luminescent molecular rods—transition-metal alkynyl complexes. *Top Curr Chem.* 2005;257:1–32. (g) Hoeben FJM, Jonkheijm P, Meijer EW, Schenning APHJ. About supramolecular assemblies of π-conjugated systems. *Chem Rev.* 2005;105:1491–1546. (h) Lu W, Chan MCW, Zhu NY, Che CM, He Z, Wong KY. Structural basis for vapoluminescent organoplatinum materials derived from noncovalent interactions as recognition components. *Chem Eur J.* 2003;9:6155–6166. (i) UHF B. Poly(aryleneethynylene)s: syntheses, properties, structures, and applications. *Chem Rev.* 2000;100:1605–1644. (j) Hide F, DiazGarcia MA, Schwartz BJ, Heeger AJ. New developments in the photonic applications of conjugated polymers. *Acc Chem Res.* 1997;30:430–436.
29. (a) Tang CW, Vanslyke SA. Organic electroluminescent diodes. *Appl Phys Lett.* 1987;51:913–915. (b) Burroughes JH, Bradley DDC, Brown AR, et al. Light-emitting diodes based on conjugated polymers. *Nature.* 1990;347:539–541. (c) Luh TY, Basuand S, Chen RN. Electroluminescent polymeric materials. *Curr Sci.* 2000;78:1352–1357. (d) Kulkarni AP, Tonzola CJ, Babel A, Jenekhe SA. Electron transport materials for organic light-emitting diodes. *Chem Mater.* 2004;16:4556–4573. (e) D"Andrade BW, Forrest SR. White organic light-emitting devices for solid-state lighting. *Adv Mater.* 2004;16:1585–1595.
30. (a) Yang J-S, Yan J-L. Central-ring functionalization and application of the rigid, aromatic, and H-shaped pentiptycene scaffold. *Chem Commun.* 2008;1501–1512. (b) Lee Y-T, Chiang C-L, Chen C-T. Solid-state highly fluorescent diphenylaminospirobifluorenylfumaronitrile red emitters for non-doped organic light-emitting diodes. *Chem Commun.* 2008;217–219. (c) Wang J, Zhao DYC, Sun H, et al. Alkyl and dendron substituted quinacridones: synthesis, structures, and luminescent properties. *J Phys Chem B.* 2007;111:5082–5089. (d) Wu C-W, Tsai C-M, Lin H-C. Synthesis and characterization of poly(fluorene)-based copolymers containing various 1,3,4-oxadiazole dendritic pendants. *Macromolecules.* 2006;39:4298–4305. (e) Lim S-F, Friend RH, Rees ID, et al. Suppression of green emission in a new class of blue-emitting polyfluorene copolymers with twisted biphenyl moieties. *Adv Funct Mater.* 2005;15:981–988. (f) He F, Tang Y, Wang S, Li Y, Zhu D. Fluorescent amplifying recognition for DNA G-quadruplex folding with a cationic conjugated polymer: a platform for homogeneous potassium detection. *J Am Chem Soc.* 2005;127:12343–12346. (g) Fan C, Wang S, Hong JW, Bazan GC, Plaxco KW, Heeger AJ. Beyond superquenching: hyper-efficient energy transfer from conjugated polymers to gold nanoparticles. *Proc Natl Acad Sci USA.* 2003;100:6297–6301. (h) Setayesh S, Grimsdale AC, Weil T, et al. Polyfluorenes with polyphenylene dendron side chains: toward non-aggregating, light-emitting polymers. *J Am Chem Soc.* 2001;123:946–953. (i) Hecht S, Frechet JMJ. Dendritic encapsulation of function: applying nature's site isolation principle from biomimetics to materials science. *Angew Chem, Int Ed.* 2001;40:74–91. (j) Jakubiak R, Bao Z, Rothberg L. Dendritic sidegroups as three-dimensional barriers to aggregation quenching of conjugated polymer fluorescence. *Synth Met.* 2000;114:61–64. (k) Kraft A, Grimsdale AC, Holmes AB. Electroluminescent conjugated polymers-seeing polymers in a new light. *Angew Chem, Int Ed.* 1998;37:402–428.

31. (a) Nguyen BT, Gautrot JE, Ji C, Brunner P-L, Nguyen MT, Zhu XX. Enhancing the photoluminescence intensity of conjugated polycationic polymers by using quantum dots as antiaggregation reagents. *Langmuir*. 2006;22:4799–4803. (b) Kulkarni AP, Jenekhe SA. Blue light-emitting diodes with good spectral stability based on blends of poly(9,9-dioctylfluorene): interplay between morphology, photophysics, and device performance. *Macromolecules*. 2003;36:5285–5296. (c) Gaylord BS, Wang S, Heeger AJ, Bazan GC. Water-soluble conjugated oligomers: effect of chain length and aggregation on photoluminescence-quenching efficiencies. *J Am Chem Soc*. 2001; 123:6417–6418. (d) Chen L, Xu S, McBranch D, Whitten D. Tuning the properties of conjugated polyelectrolytes through surfactant complexation. *J Am Chem Soc*. 2000;122:9302–9303. (e) Taylor PN, MJ O"C, McNeill LA, Hall MJ, Aplin RT, Anderson HL. Insulated molecular wires: synthesis of conjugated polyrotaxanes by Suzuki coupling in water. We are grateful to Carol A Stanier for valuable discussion and to Professor Christopher J. Schofield for providing facilities for gel electrophoresis. Disodium 1-aminonaphthalene-3,6-disulfonate was generously provided by Dr. M. G. Hutchings of BASF plc (Cheadle Hulme, UK). This project is funded by the Engineering and Physical Sciences Research Council (UK). *Angew Chem, Int Ed*. 2000;39:3456–3460. (f) Sainova D, Miteva T, Nothofer G, et al. Control of color and efficiency of light-emitting diodes based on polyfluorenes blended with hole-transportingmolecules. *Appl Phys Lett*. 2000;76:1810–1812.
32. (a) Luo J, Xie Z, Lam JWY, et al. Aggregation-induced emission of 1-methyl-1, 2,3,4,5-pentaphenylsilole. *Chem Commun*. 2001;1740–1741. (b) Tang BZ, Zhan X, Yu G, Lee PPS, Liu Y, Zhu D. Efficient blue emission from siloles. *J Mater Chem*. 2001;11:2974–2978.
33. Freemantle M. New horizons for ionic liquids. *Chem Eng News*. 2001;79:21–25.
34. (a) Mei J, Leung NLC, Kwok RTK, Lam JWY, Tang BZ. Aggregation-induced emission: together we shine, united we soar! *Chem Rev*. 2015;115:11718–11940. (b) Hong Y, Lam JWY, Tang BZ. Aggregation-induced emission. *Chem Soc Rev*. 2011;40: 5361–5388.
35. Zeng Q, Li Z, Dong Y, et al. Fluorescence enhancements of benzene-cored luminophors by restricted intramolecular rotations: AIE and AIEE effects. *Chem Commun*. 2007;70–72.
36. (a) Deng C, Tang L, Qin A, Hu R, Sun JZ, Tang BZ. Specific detection of D-glucose by a tetraphenylethene-based fluorescent sensor. *J Am Chem Soc*. 2011;133:660–663. (b) Liu L, Zhang G, Xiang J, Zhang D, Zhu D. Fluorescence "turn on" chemosensors for Ag^+ and Hg^{2+} based on tetraphenylethylene motif featuring adenine and thymine moieties. *Org Lett*. 2008;(20):4581–4584. (c) Wang M, Zhang G, Zhang D, Zhu D, Tang BZ. Fluorescent bio/chemosensors based on silole and tetraphenylethene luminogens with aggregation-induced emission feature. *J Mater Chem*. 2010;20:1858–1867. (d) Yuan WZ, Gong Y, Chen S, et al. Efficient solid emitters with aggregation-induced emission and intramolecular charge transfer characteristics: molecular design, synthesis, photophysical behaviors, and OLED application. *Chem Mater*. 2012;24:1518–1528. (e) Khandare DG, Kumar V, Chattopadhyay A, Banerjee M, Chatterjee A. An aggregation-induced emission based "turn-on" fluorescent chemodosimeter for the selective detection of ascorbate ions. *RSC Adv*. 2013;3:16981–16985. (f) Shi H, Kwok RTK, Liu J, Xing B, Tang BZ, Liu B. Real-time monitoring of cell apoptosis and drug screening using fluorescent light-up probe with aggregation-induced emission characteristics. *J Am Chem Soc*. 2012;134:17972–17981. (g) Yang X, Shen B, Jiang Y, et al. A novel fluorescent polymer brushes film as a device for ultrasensitive detection of TNT. *J Mater Chem A*. 2013;1:1201–1206. (h) Khandare DG, Joshi H, Banerjee M, Majik MS, Chatterjee A. An aggregation-induced emission based "turn-on" fluorescent chemodosimeter for the selective detection of Pb^{2+} ions. *RSC Adv*.

2014;4:47076–47080. (i) Wu Y, Hu J, Huang H, et al. Memory chromic polyurethane with tetraphenylethylene. *J Polym Sci, Part B: Polym Phys*. 2014;52:104–110. (j) Carbas BB, Odabas S, Turksoy F, Tanyeli C. Synthesis of a new electrochromic polymer based on tetraphenylethylene cored tetrakis carbazole complex and its electrochromic device application. *Electrochim Acta*. 2016;193:72–79. (k) Khandare DG, Joshi H, Banerjee M, Majik MS, Chatterjee A. Fluorescence turn-on chemosensor for the detection of dissolved CO_2 based on ion-induced aggregation of tetraphenylethylene derivative. *Anal Chem*. 2015;87:10871–10877. (l) Chatterjee A, Khandare DG, Saini P, Chattopadhyay A, Majik MS, Banerjee M. Amine functionalized tetraphenylethylene: a novel aggregation-induced emission based fluorescent chemodosimeter for nitrite and nitrate ions. *RSC Adv*. 2015;5:31479–31484.
37. Li Y, Wu Y, Chang J, Chen M, Liu R, Li F. A bioprobe based on aggregation induced emission (AIE) for cell membrane tracking. *Chem Commun*. 2013;49:11335–11337.
38. Liu D, Ji S, Li H, et al. Cellular membrane-anchored fluorescent probe with aggregation-induced emission characteristics for selective detection of Cu^{2+} ions. *Faraday Discuss*. 2017;196:377–393.
39. Zhang W, Yu CYY, Kwok RTK, Lamab JWY, Tang BZ. A photostable AIE luminogen with near infrared emission for monitoring morphological change of plasma membrane. *J Mater Chem B*. 2018;6:1501–1507.
40. Zhang Y, Yan Y, Xia S, et al. Cell membrane-specific fluorescent probe featuring dual and aggregation-induced emissions. *ACS Appl Mater Interfaces*. 2020;12:20172–20179.

CHAPTER FIVE

Aggregation-induced emission luminogens for lipid droplet imaging

A.H.M. Mohsinul Reza[a,b], Yabin Zhou[a], Jianguang Qin[a,*], and Youhong Tang[a,b,*]

[a]College of Science and Engineering, Flinders University, Adelaide, SA, Australia
[b]Institute for NanoScale Science and Technology, College of Science and Engineering, Flinders University, Adelaide, SA, Australia
*Corresponding authors: e-mail address: jian.qin@flinders.edu.au; youhong.tang@flinders.edu.au

Contents

1. Introduction — 102
2. Significance of lipid drops study in different organisms — 103
    2.1 Disease associated with lipid disorders in human — 104
    2.2 Lipid drops from microorganisms for human benefits — 104
3. Biosynthesis of lipid drops — 106
4. Advancement and challenges in lipid research — 108
    4.1 Common methods applied in modern lipidomics analysis — 110
    4.2 Advantages and disadvantages of the existing probes for lipid detection — 110
5. Recent progress of lipid-specific probes with aggregation-induced emission — 113
    5.1 Photostable lipid-specific probes based on simple AIEgens — 114
    5.2 Two-photon lipid droplets specific AIE bioprobes — 116
    5.3 AIE-based fluorophore to visualize lipid drops-lysosome interplay — 118
    5.4 Theranostics approaches with lipid-specific AIE probes — 120
    5.5 Lipid-specific AIEgens system with wide emission tunability — 122
    5.6 Biocompatible AIEgen from natural resources — 123
    5.7 Lipid specific AIEgens in algae research — 126
6. Conclusions — 138
7. Future remarks — 138
Acknowledgments — 138
References — 138

Abstract

Lipid droplets (LDs) are evolutionarily conserved organelles involved in energy homeostasis and versatile intracellular processes in different cell types. Their importance is ubiquitous, ranges from utilization as the biofunctional components to third-generation biofuel production from microalgae, while morphology and functional perturbations could also relate to the multiple diseases in higher mammals. Biosynthesis of lipids can be triggered by multiple factors related to organismal physiology and the surrounding environment. An early prediction of this might help take necessary actions toward desired outcomes. In vivo visualization of LDs can give molecular insight into regulatory mechanisms and the underlying connections with other cellular structures. Traditional bioprobes for LDs detection often suffer from different dye-specific limitations such as aggregation-caused quenching and self-decomposition phenomena that hinder the research advancement. The emergence of lipid-specific nanoprobes with aggregation-induced emission (AIE) attributes in recent years is promising in remunerative characteristics with defined bioimaging properties. By utilizing the easy synthetic techniques and exploiting the unique physical features of these molecules, highly selective, stable, biocompatible and facile fluorescent probes could be fabricated for lipid detection. This chapter will provide up-to-date insight into the recent advances in lipid-specific AIE-based probes to enhance the opportunities for basic research related to the distinct roles of LDs in living organisms.

Graphical abstract

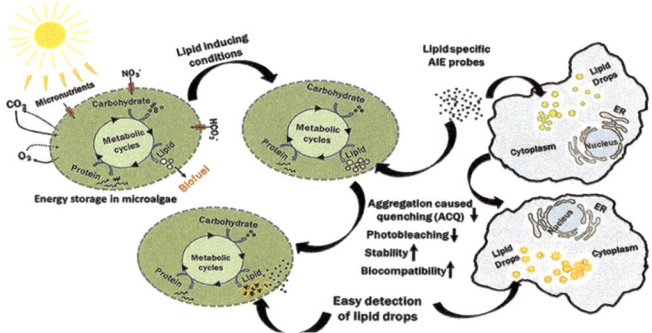

Introduction of aggregation-induced emission (AIE) luminogens for lipid droplet imaging.

1. Introduction

Lipid droplets (LDs) are evolutionarily conserved structures in prokaryotic and eukaryotic cells and have long been focused on complex and multiplexed biological functions.[1,2] These unique organelles are the deposited forms of neutral lipids, such as triacylglycerides (TAG) and sterol esters (SE).[3] Generally, LDs are coated with structural proteins and phospholipid monolayer, and stored to be utilized as the metabolic energy source at the time of necessities.[4] However, their structural organization has been debatable from the analysis of immunofluorescence microscopy, live-cell imaging with fluorescently tagged LD surface proteins and the results from the electron microscopy. It happened possibly due to the differences in sample preparation techniques.[5,6] The synchronous participation of LDs with other cellular components,[7,8] distinctive architectures and inadequacy of stainable materials can also trigger dehydration and lack of LDs in histological research and makes the study of lipid biology more challenging in a biological system.[9]

This chapter aims to provide up-to-date insight into the multifaceted applications of these AIE molecules with the expectation to broaden the opportunities through the proper selection of these AIE fluorophores. This chapter comprises four major sections. We first outline the significance of LDs study in different organisms and then discuss biosynthesis of lipid drops. Subsequently, we further update the advancement and challenges in lipid research and then summarize the recent progress of lipid-specific probes with aggregation-induced emission. In the conclusion section, we point out a new research area on the use of AIE materials for lipid droplet imaging and detection in microalgae as there is a great potential to use microalgae as a microfactory to produce biofuel or healthy food supplements for human.

2. Significance of lipid drops study in different organisms

The previously conceived LDs as the cytoplasmic annexation of fat[10] is now considered independent organelles with key regulatory roles in energy homeostasis and signaling in multiple organisms, ranging from mammals to plants.[4,11] Recent proteomic advances suggest the direct involvement of several groups of functional proteins in lipid biosynthesis that differs in cell types.[12] So far perceived, the main function of LDs is to store energy mainly in the form of highly condensed and compact structures of triacylglycerols in organisms and facilitate surplus energy during adverse and nutrient deficient environmental conditions.[11] From different studies, the possibilities of changes in buoyancy of organisms to facilitate them switching the adverse environment toward favorable conditions is considered. The development of diatom-derived polyunsaturated aldehydes (PUAs) and oxylipins based defense system have also been predicted in microorganisms that are originated from polyunsaturated fatty acids (PUFAs).[13–15] Other complex involvement of LDs in several physiological functions has also been reported. For example, in the developmental period of *Drosophila*, prior to the inclusion in the dividing nuclei, accumulation of histones at LDs surface has been studied, and ontogenetic relationship among host lipid droplets and different intracellular pathogens like hepatitis C virus, dengue virus and rotaviruses have been observed.[16,17]

2.1 Disease associated with lipid disorders in human

Functional deregulations of LDs have been associated with several human diseases. Excessive accumulation of LDs could engender major health issues like obesity, liver steatosis, atherosclerosis, cardiac dysfunctions and type 2 diabetes mellitus.[18] On the other hand, a heterogeneous group of diseases that are often termed as lipodystrophy syndromes caused by the selective absence of adipose tissues leads to insulin resistance and other metabolic complications in human.[19] Some inherited disorders caused by the mutations in adipose triglyceride lipase and metabolic gene expression results in excess accumulation of TAG-rich cytosolic LDs in important organelles that cause neutral lipid storage diseases.[20] LDs also affect other physiological processes of the central nervous systems beyond their active roles of energy processing. Being the second most lipid-rich organ, neurodegenerative diseases like epilepsy, Alzheimer's disease, schizophrenia, Parkinson's disease, Huntington's disease and motor neuron diseases have been characterized by cytotoxic consequences of lipid dysfunction.[21] Additionally, compelling evidence also suggests critical linkage among malfunctioning LDs homeostasis, cell signaling, cancer cell proliferation and aggravation.[22]

2.2 Lipid drops from microorganisms for human benefits

Over the last few decades, utilization of plant-based biofuels has become controversial due to the food-fuel conflicts. Processing biofuels from plant-based sources is also expensive, and most often requires using toxic chemicals that are environmental hazards.[23] Therefore, over the period, oleaginous microorganisms like yeast, bacteria and microalgae have gained industrial attention as the substantial sources of biofuel owing to their advantages such as fast growth rates, higher oil yielding capability, the usability of wastewaters and non-arable land, and procurement of high-valued biofunctional compounds.[24–26] Generally, triacylglycerols (TAGs), wax esters (WEs), polyhydroxyalkanoates (PHAs) and polyhydroxybutyrate (PHB) from these organisms could be utilized as the lipid-based biofuel sources. As an excellent genetic model organism yeast is used for dynamic LDs research. Nearly all yeast can produce lipids. However, some yeasts, such as *Yarrowia lipolytica* and *Saccharomyces cerevisiae* can produce TAGs and WEs in a large amount, have been bioengineered to produce PHA/PHB.[27,28] Some gram-positive bacteria like *Nocardia*, *Mycobacterium*, *Rhodococcus*, *Dietzia*, and *Streptomyces*, and gram-negative bacteria like *Marinobacter*, *Pseudomonas* sp., and *Acinetobacter* can also produce TAG and WEs under nitrogen starved and carbon supplemented conditions.[29]

As the alternative resources of biofuel, microalgae have great potential with extenuating advantages of the greenhouse gas emission, high biomass and conversion efficiency of solar to chemical energy.[30] Several microalgal strains, such as *Scenedesmus obliquus*,[31] *Dunaliella tertiolecta*,[32] *Chlorella vulgaris*, *Spirulina platensis*,[33] *Chlorococcum parinum*, *Chlorococcum littorale*,[34] *Nannochloropsis oculata*,[35] and *Haematococcus pluvialis*[36] have been identified as a potential source of TAG. Additionally, *E. gracilis* is an excellent source of myristic acid (C14:0) that has a lower freezing point and a good cetane number (66.2), therefore, can be utilized for drop-in jet fuel. The extent of unsaturation of other PUFAs with more than four double bonds can be reduced by partial catalytic hydrogenation of the oil.[37,38] The fuel properties can also be improved further by modifying the fatty ester composition for direct combustion in sensitive engines.[39]

The metabolites procured from these microorganisms are also important for different pharmaceutical aspects. Due to the high lipid production capabilities and fast growth rates, valuable dietary constituents of antioxidants, PUFAs, vitamins and other health effective bioactive compounds can be achieved.[40] Dietary inclusion of LC-PUFAs has long been signified and associated with the lower risk factors of coronary heart disease and stroke.[41,42] By balancing the ω-6/ω-3 ratio in human, this could also aid cancer and neurodegenerative diseases by cutting down the risk factors these diseases.[43,44] As the building block precursors of distinguished prostaglandins, leukotrienes, lipoxines, and, tromboxanes, arachidonic acid (C20:4) and eicosapentaenoic acid (20:5, n-3) can be utilized as medicaments in conditions like asthma and atopic diseases.[45,46] Palmitoleic acid (16:1n-7) consumption can also ameliorate obesity, diabetes and hepatosteatosis,[47,48] while omega-6 polyunsaturated fatty acids and linoleic acid (18:2n-6) can reduce hair loss and assist with wound healing.[49,50] DHA (22:6n-3) can also boost visual activities, and improve cognitive functions and immunity in human through the multiplexed cell membrane and cell signaling.[51–53]

3. Biosynthesis of lipid drops

Despite the advancement in biological research and numerous studies focusing on LDs, the fundamental questions related to the biogenesis of LDs, regulation mechanisms and biological fates are not clearly understood. In higher organisms, LDs structures provide unique separation of the aqueous and organic phases of the cell, where involvement of several structural proteins, e.g., proteins of the perilipin family,[54] lipid synthesis enzymes, e.g., acetyl coenzyme A (CoA) carboxylase, acyl-CoA synthetase, acyl-CoA: diacylglycerol acyltransferase 2 (DGAT2), lipases, e.g., adipose tissue triacylglycerol lipase (ATGL) and membrane-trafficking proteins, e.g., Rab5, Rab18 and ARF1 have been observed.[55,56] However, the regulatory mechanisms and localization of target proteins are still obscure. Among different models of the LDs biosynthesis[11] in mammals, the model of Tan et al.[57] is considered most acceptable (Fig. 1).[58] According to the model, lipid bilayer vesicles form due to the accumulation of triglyceride (TG) in the endoplasmic reticulum (ER) membranes and grow to the primary LDs with the addition of more phospholipids to it. Enzymes required for the LDs formation, stabilization, and degradation are also located in the ER. Subsequently, secondary LDs form and enlarge to matured LDs through the fusion of the primary LDs under tight regulation of several functional proteins located on the surface and nearby the LDs. Interactions among these proteins, adipose triglyceride lipase (ATGL) and respective genes/proteins regulate the levels of intracellular triglyceride. The matured LDs are coated by a membrane monolayer of phospholipids and sphingomyelin, cored with TG and cholesterol esters (Fig. 1A). Hydrolysis of LDs is further synchronized by the consecutive actions of ATGL, hormone-sensitive lipase (HSL), and monoglyceride lipase (MGL) that act through hydrolyzing the triglyceride, cleaving one molecule of fatty acid from diacylglycerol, and through final hydrolysis by MGL, respectively. In hepatocytes that carry out many critical metabolic functions, the lipolysis reduces during the interaction of ATGL with LD protein, perilipin 5 (PLIN5) (Fig. 1B), and the process is upsurged while ATGL interacts with and comparative gene identification-58 (CGI-58).[59]

In microorganisms, the lipid metabolic pathways are also branched out due to the evolutionarily diversified organisms. However, using radioactive tracer and ultrastructural analysis, as well from the transcriptome, proteome and metabolome profiles of different fungal and higher plants,[60,61] in green

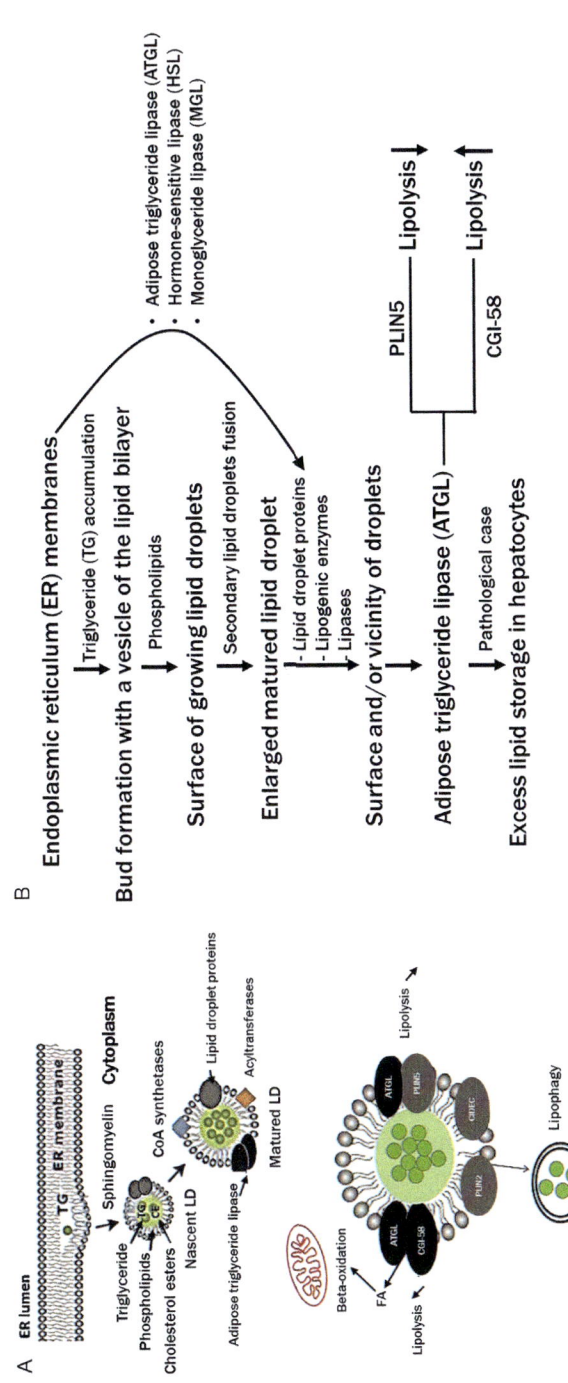

Fig. 1 Lipid droplet biology. (A) Formation of lipid droplet (LD) with lipid droplet proteins and lipolysis and (B) flow diagram of LDs biosynthesis, hydrolysis and storage in hepatocytes.[58] ER, endoplasmic reticulum; FA, fatty acid; TG, triglyceride; ATGL, adipose triglyceride lipase; HSL, hormone-sensitive lipase; MGL, monoglyceride lipase; PLIN2, LD protein, perilipin 2; PLIN5, LD protein, perilipin 5; CIDEC, LD protein, CIDEC; CGI-58, comparative gene identification-58.

algae pathways for the lipid biosynthesis have been proposed.[15,61–63] Briefly, both the endoplasmic reticulum-derived compartments and the chloroplasts are involved in TAG synthesis. These are also tightly regulated by different acyltransferase isoforms.[61,64] The step starts in the chloroplast with the subsequent conversion of acetyl-CoA to malonyl-CoA and malonyl acyl carrier proteins (malonyl-ACP), catalyzed by acetyl-CoA carboxylase (ACCase) and malonyl-CoA transacylase, respectively. The lengthening of the acyl chains from the substrates malonyl-ACP and acetyl-CoA is catalyzed by fatty acid synthase (FAS), which consequently hydrolyze to free fatty acids (FFA) through the actions of fatty-ACP thioesterases (FAT). These fatty acids (FAs) could be used as the building blocks of diacylglycerol (DAG) and chloroplastic TAG synthesis through the acylation of glycerol-3-phosphate (G3P) to serve the photosynthetic membrane lipids.[64] The TAG biosynthesis pathway involves enzymes glycerol-3-phosphate acyltransferase (GPAT), lysophosphatidate acyltransferase and phosphatidic acid phosphatase (PAP) for consecutive catalysis of G3P into lysophosphatidate (LPA), phosphatidate (PA) and DAG, respectively. The catabolic products of FAs can also be transported to cytosol and utilize coenzyme A and long-chain acyl-CoA synthetase (LACS) to synthesis acyl-CoA for production of more TAG through successive elongation and desaturation in the endoplasmic reticulum (Fig. 2).[15]

4. Advancement and challenges in lipid research

The accelerated progresses in LD research are currently mostly focused on animal models to apprehend the decisive roles of LDs in prominent diseases, and triggering their synthesis in photosynthetic organisms as the beneficiary biofunctional products. Depending on distinct factors, lipid biosynthesis in organisms varies significantly. Nutrient composition, environmental dynamics, genetic and hormonal functioning are the key influencers, where large networks of genes and proteins are involved in modeling lipid production. Impairment of the LDs biosynthesis machinery in cells and deregulations can result in severe physiological anomalies in higher organisms.[65,66] On the other hand, different strategies to trigger lipid biosynthesis in microorganisms are being studied for human benefits. Most often in microalgae, nutrient-enriched conditions direct the carbon flux toward the biosynthesis of macromolecules and other cellular constituents.[67,68] This could be altered to the synthesis of non-N containing

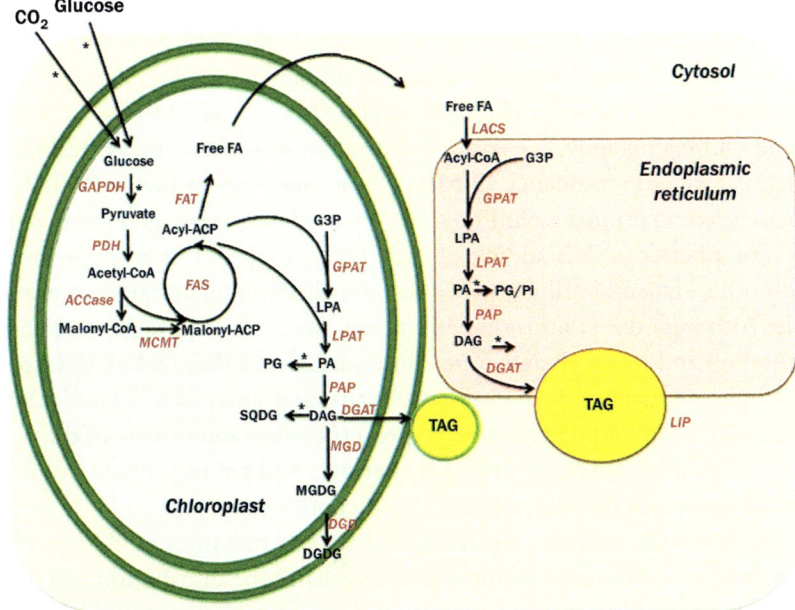

Fig. 2 Triacylglycerol (TAG) biosynthesis de novo in microalgae. The enzymes are shown in orange. AADA, alpha amylase domain-containing protein; ACCase, acetyl-CoA carboxylase; ACP, acyl carrier protein; CoA, coenzyme A; DAG, diacylglycerol; DGAT, diacylglycerol acyltransferase; DGD, digalactosyldiacylglycerol synthase; FAT, fatty acyl-ACP thioesterase; FAS, fatty acid synthase; free FA, free fatty acid; G3P, glycerol 3-phosphate; GPAT, glycerol 3-phosphate acyltransferase; LACS, long-chain acyl-CoA synthetase; LPA, lysophosphatidic acid; LPAT, lysophosphatidic acid acyltransferase; MCMT, malonyl-CoA:acyl carrier protein malonyltransferase; MGD, monogalactosyldiacylglycerol synthase; PA, phosphatidic acid; PDAT, phospholipid: diacylglycerol acyltransferase; PG, phosphatidylglycerol; SQDG, sulfoquinovosyl diacylglycerol; PI, phosphatidylinositol; PE, phosphatidylethanoalamine; PDH, pyruvate dehydrogenase; PAP, phosphatidic acid phosphatase; TAG, triacylglycerol. *Some detailed reactions are excluded to simplify the illustration.[15,61–63]

compounds like lipids through the utilization of the available cellular carbon precursors by exposing them to the nitrogen deplete conditions.[15,69,70] Combinations of different nutrient starvation, notably sulfur, iron, phosphorus and potassium have also been studied to increase lipid biosynthesis in microalgae.[71–73] However, the applicability of these results is still contingent upon the exploration of the interacted signaling pathways, a large number of genes, and growth factors involved in lipid biology.[62,74]

4.1 Common methods applied in modern lipidomics analysis

Despite the challenges of lipidomics analysis due to the lipids diversity,[75] recent advancements in lipid research herald significant progress in the analytical chemistry area. Thin-layer chromatography (TLC),[76] gas chromatography,[77] liquid chromatography,[78] enzyme-linked immunosorbent assays (ELISA),[79] nuclear magnetic resonance (NMR),[80] and mass spectrometry (MS)[81] are the most conventional techniques, widely used to understand lipid dynamics. For a better understanding of the LDs biology and their involvement with other dynamic cellular processes, visualization techniques are preferable. Although direct infusion MS strategy does not require previous lipid separation and relatively less time consuming than other techniques, spectroscopic approaches do not allow visualization analysis of LDs.[75] Other methods require biological extracts, pre-processing and storage of the samples[82,83] that are time-consuming, expensive, and fail to provide information about spatial distribution in in vivo conditions.

In nanoscale analysis, electron microscopy and immunofluorescence microscopy are commonly employed to visualize the distribution and morphology of LDs.[84] As the non-invasive and non-destructive method, Raman microscopy,[85] coherent anti-Stokes Raman scattering microscopy,[86] and direct organelle MS[87] are also in current practices. These technologies allow studying the biophysics of LDs, but the requirement of sophisticated equipment, complex data processing and sample preparations often impede the proper objectives. Sometimes, cells must be fixed, or LDs need extraction for further analysis, sacrificing the opportunity to detect the simultaneous activities and dynamics of LDs in integrity.

4.2 Advantages and disadvantages of the existing probes for lipid detection

Microscopy has shaped our comprehension of fundamental research in cells as living units. To interpret dynamic functions and biological interactions at the molecular level, microscopic techniques are still central. Advancement in fluorescence microscopy techniques and its large temporal resolution allow us new perceptive to access sensitive information of lipid organization. Availability of numbers of lipid binding probes and fluorophore-conjugated lipids are now opening up possibilities for detailed imaging of lipid drops and intracellular compartments for a superior understanding of the biophysical properties of lipid biology.[88–90] Therefore, the new probes are superior to other techniques for a rapid study of LDs dynamics.[91–93]

Currently, among the lipid-specific traditional dyes, Sudan Black B ((2,2-dimethyl-1,3-dihydroperimidin-6-yl)-(4-phenylazo-1-naphthyl)diazene),[94] Sudan III (1-(4-(phenyldiazenyl)phenyl)azonaphthalen-2-ol),[95] Oil Red O (1-(2,5-dimethyl-4-(2,5-dimethylphenyl)phenyldiazenyl)azonapthalen-2-ol))[96] are used under bright-field microscopy. The sample preparation techniques with these dyes are time-consuming and need ethanol or isopropanol for extraction. Varying filtration conditions and alterations of LDs morphologies during fixative stages often lead to inconsistent results and are major constraints for utilizing these dyes.[97,98]

Nile Red (9-diethylamino-5H-benzo[a]phenoxazine-5-one),[99] BODIPY 493/503 (4,4-difluoro-1,3,5,7,8-pentamethyl-4-bora-3a,4a-diaza-s-indacene) and BODIPY 505/515 (4,4-difluoro-1,3,5,7-tetramethyl-4-bora-3a,4a-diaza-s-indacene)[100,101] (Thermo Fisher Scientific BODIPY 505/515, 2019; Thermo Fisher Scientific BODIPY 493/503, 2019) are well recognized and extensively used commercial fluorophores for fluorescent microscopy. Depending on environment polarity, the emission of Nile Red varies from the color of deep red to strong yellow-gold (yellow-gold fluorescence: $\lambda_{ex} = 450–500$ nm and $\lambda_{em} = \geq 528$ nm; red fluorescence: $\lambda_{ex} = 515–560$ nm and $\lambda_{em} = \geq 590$ nm).[99] A decrease in the surrounding polarity might result in a blue-shifted emission peak for Nile Red, while the broad absorption and emission spectra can cause a signal overlap between the yellow and red spectra.[102,103] These properties also make the dye challenging to utilize for multicolor imaging purposes. Additionally, other cellular structures with protein and nonlipid in the hydrophobic domains are often reported to affect the studies with Nile Red.[104] Comparing to Nile Red, BODIPY 493/503 and BODIPY 505/515 are less affected by the surrounding polarity and pH, therefore more suitable for LDs studies. However, sometimes BODIPY can bind unselectively with the mitochondrial, plasma, and nuclear membranes.[93] Additionally, dye acquisition with BODIPY and Nile Red needs extreme caution as lower concentrations can cause photobleaching, and a slightly higher concentration or lower level of competencies during washing steps could engender marked background signals and results in ineffective lipid studies.[105,106] Table 1 shows common lipid-specific dyes for lipid detection.

Table 1 Properties of Some Commercially Available Lipid-Specific Dye.[14]

Common Dyes for Lipid Detection in Bright-Field Microscopy

Indicator	Chemical structure	Solubility	Notes
Sudan Black B[94]	Chemical formula: $C_{29}H_{24}N_6$ MW: 456.553	C_6H_5OH, C_2H_5OH, C_3H_6O, C_6H_6, C_7H_8	• Nonfluorescent • Can stain other materials, but not so lipid specific • Distinguish hematological disorders and fingerprint enhancement
Sudan III[95]	Chemical formula: $C_{22}H_{16}N_4O$ MW: 352.39	$CHCl_3$, C_7H_8 (1 mg/mL), H_2O (<0.1 mg/mL), C_2H_5OH (2 mg/mL), C_3H_8O (saturated solutions)	• Emit red fluorescence • Low solubility, requires dissolving in ethanol or isopropanol • Sometimes cause disruption and fusion of the lipid drops
Oil Red O[96]	Chemical formula: $C_{26}H_{24}N_4O$ MW: 408.49	C_2H_5OH: $CHCl_3$ (1:1)	• Emit red fluorescence for fluorescence microscopy • Low solubility, requires solvent like ethanol • Might cause the disruption and fusion of lipid drops

Commercial dyes for fluorescent microscopy

Indicator	Chemical structure	$\lambda_{ex}/\lambda_{em}$ (nm)	Solubility	Notes
Nile Red[99]	Chemical formula: $C_{20}H_{18}N_2O_2$ MW: 318.37	λ_{ex}: 450–500; λ_{em}: >528; λ_{ex}: 515–560, λ_{em}: >590	MeOH, C_2H_5OH, $C_3H_8O_2$, C_3H_7NO, DMSO	• Wide absorption and emission spectra, unsuitable for multicolor imaging • Highly sensitive to the environmental polarity • Yellow-gold fluorescence in nonpolar condition; red fluorescence in polar condition

Table 1 Properties of Some Commercially Available Lipid-Specific Dye.[14]—cont'd

			Commercial dyes for fluorescent microscopy	
Indicator	Chemical structure	$\lambda_{ex}/\lambda_{em}$ (nm)	Solubility	Notes
				• Photobleaching might cause by self-decomposition
				• Nonselective binding to cellular structures with a hydrophobic domain
BODIPY 493/503[100]	![structure]	λ_{ex}: 493; λ_{em}: 503	C_2H_5OH, DMSO	• Self-quenching fluorescence diminishing for aggregation at higher concentrations
	Chemical formula: $C_{14}H_{17}BF_2N_2$ MW: 262.1085			• More specific for cellular lipid droplets than Nile Red
BODIPY 505/515[101]	![structure]	λ_{ex}: 505; λ_{em}: 515	C_2H_5OH, DMSO	• Self-quenching fluorescence diminishing for aggregation at higher concentrations
	Chemical formula: $C_{13}H_{15}BF_2N_2$ MW: 248.0817			• More specific for cellular lipid droplets than Nile Red

5. Recent progress of lipid-specific probes with aggregation-induced emission

Most of the existing dyes for LDs detection may suffer from aggregation caused quenching (ACQ) and reduce the fluorescence properties in aggregated state at high concentration due to π-π stacking.[107] In recent years, the use of aggregation-induced emission (AIE), the opposite phenomena of ACQ of some molecules, and high-quality biocompatible fluorophores has been introduced in molecular and biological studies.[108] Through binding with the targeted biomolecules or influenced by the surrounding environment, the intramolecular motions in these molecules restrict and exhibit increased emission. The incorporation of AIEgen nanoparticles also possesses improved photostability and makes them suitable for long-term high contrast imaging in a biological environment.[107] Due to the higher fraction of triglycerides and cholesterol esters, the inherent hydrophobic environment of lipid droplets make them more potential for staining with AIEgens. Over the last few years, numbers of lipid-specific AIE probes with a diverse range of excitation and emission wavelengths have been synthesized, each of which has unique properties and could be selected effectually based on study requirements. Among these AIE probes, some follow one-pot synthetic strategies that allow the synthesis of other fluorophores in time and cost-effective manners. Some of these probes could be excited at near-infrared light and suitable for deep tissue penetration, and some have already shown their significant therapeutic properties. In the following section, different LDs-specific AIE-based fluorophores with their unique properties have been discussed that might be useful for the potential users to find suitable probes during studies.

5.1 Photostable lipid-specific probes based on simple AIEgens

Applications of AIEgens in lipidomics have been highly appreciated since the biosynthesis of lipid-specific AIEgens, such as FAS ($C_{20}H_{14}N_2O$) and DPAS ($C_{20}H_{16}N_2O$) are economical, and produced from low-cost commercial products like fluorenone and benzophenone, respectively.[109] Previously, tetraphenylethylene (TPE) derivatives, TPE-AmAl and TPE-AC have been synthesized as bright and photostable lipid-specific AIEgens with excellent cell permeability by Wang et al.[110] and Kang et al.,[111] respectively (Fig. 3A and B). The fluorescence of TPE-AmAl is responsive to environmental polarity. With an increase in water fraction (from 0 to 70 vol%), the emission of this AIE probe has been red-shifted (525–590 nm) while excited at 330–385 nm. Further increase of water fraction has resulted in consequent blue-shifting of this fluorophore. Due to the aggregation of TPE-AmAl inside the less polar LDs it emits greenish-blue color, and could easily be distinguished from the extracellular orange emission. Besides, compared to the traditional LD-specific fluorophores, TPE-AmAl exhibited very low background noise and higher sensitivity. TPE-AmAl was further reformed into a novel near-infrared (NIR) AIEgen, TPE-AC by incorporating electron donor, dimethylamine and electron acceptor,

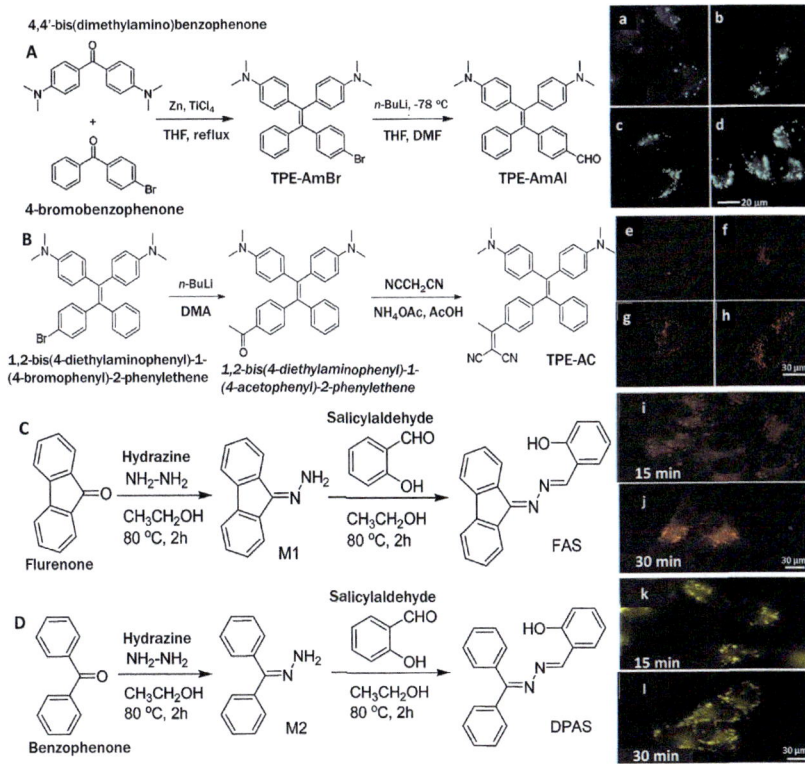

Fig. 3 Simple synthetic pathways and staining of lipid drops with AIEgens, TPE-AmAl in HeLa cells treated with oleic acid (A), TPE-AC (B), FAS (C) and DPAS (D). (a–d) HeLa cells stained with TPE-AmAl (10 μM; 15 min incubation) in the presence of (a) 0, (b) 12.5, (c) 25 and (d) 50 μM oleic acid for 6 h (λ_{ex}: 330–385 nm); (e–h) HeLa cells stained with TPE-AC (10 μM; 15 min incubation) in the presence of (a) 0, (b) 12.5, (c) 25 and (d) 50 μM oleic acid for 6 h (λ_{ex}: 510–550 nm). (i–l) HeLa cells stained with 7.5 μM FAS (λ_{ex}: 400–440 nm) (i, j) and DPAS (λ_{ex}: 330–385 nm) (k–l) for different time after incubation with oleic acid (50 μM) for 6 h. Stained time: (i, k) 15 min; (j, l) 30 min.[109–111]

malononitrile into the TPE core. In tetrahydrofuran (THF) solution, maximum absorption of this nanoprobe has been observed at 455 nm, while increasing the water fraction caused aggregation of this AIE probe and resulted in emission at 705 nm. To minimize the issues of the environmental polarity of TPE-AmAl, two AIE bioprobes, FAS and DPAS, respectively, has been synthesized by Wang et al.[109] (Fig. 3C and D). Unlike TPE-AmAl, the fluorescence of FAS and DPAS is more stable and less affected by the surrounding environment. Both the dyes showed high contrast and brightness at significantly higher concentrations, therefore, resolved the concentration-related ACQ problems of the commercial phosphorescent probes.

However, these lipid-specific AIE probes require UV excitation and mostly show short-wavelength emission. Therefore, they might damage some cellular structures during a long exposure or could be affected by some of the cellular autofluorescence properties. Additionally, for optimal tissue penetration, an excitation wavelength around 750–950 nm is preferable.[112] Despite the near-infrared (NIR) emission properties of TPE-AC, the excitation wavelength is only 450 nm, which is not long enough to serve the purposes. Given these facts, the necessities of lipid-specific two-photon lipid AIEgens have consecutively been emerged and focused on by researchers in recent years.

5.2 Two-photon lipid droplets specific AIE bioprobes

Multiphoton imaging generally allows lower background autofluorescence, minimal photobleaching, and higher spatial resolution. Owing to the high brightness and photostable characteristics, AIEgens have been considered more promising candidates for multiphoton-active imaging.[112–114] Additionally, near-infrared excitation light can facilitate lower scattering coefficient and narrower focus range, allowing deeper tissue penetration and considered more suitable for biomedical diagnosis and therapy.[115–117] Two-photon excitations in a fluorescent compound are critically determined by the molecular two-photon absorption strengths (δ_{2PA}). A higher value of δ_{2PA} shows greater strength of the fluorescence and reduced damaging effects from raising temperature due to the strong laser pulse.[118] This could provide a straightforward way to determine the red and NIR excitations. Recently, several two-photon AIE probes have been designed for fluorescence microscopy that perform better than the commercial LD staining fluorophores in terms of superior 3D resolution, minimized autofluorescence, lower photobleaching, and deeper tissue penetrability.

Among these probes, TPA-BI ($C_{32}H_{29}N_3O$)[112] has shown strong solvatochromism properties and successfully been exploited in LD imaging in several cells lines, such as A549, HeLa, and HepG-2, and in fixed cells. In comparison to the one-photon excitation (λ_{ex} = 400–440 nm), under two-photon excitation (λ_{ex} = 840, 900, and 980 nm) in HeLa cells, the response of TPA-BI to the TAG/LDs has been largely increased (λ_{em} = 450–550 nm), where BODIPY 493/503 (λ_{em} = 500–550 nm) showed significant signal loss (Fig. 4A). These properties allow the escaping of damage to the living cells due to the high laser power. Moreover, by

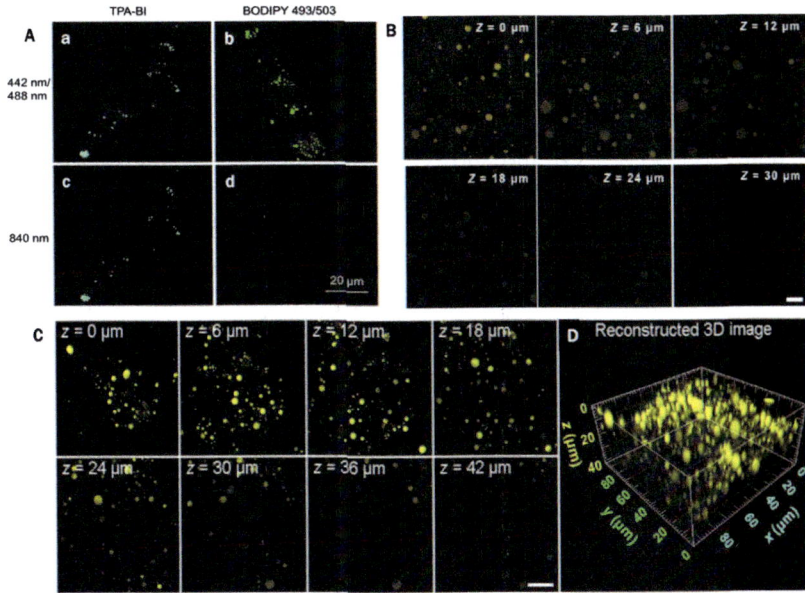

Fig. 4 Confocal fluorescence imaging with TPA-BI (A) and ABCXF (B–D). (A) HeLa cells stained with TPA-BI (5 μM) (a and c) and BODIPY 493/503 (5 μM) for 20 min (b and d). Conditions: one-photon excitation (a and b) (a) λ_{ex}: 442 nm; λ_{em}: 450–550 nm, (b) λ_{ex}: 488 nm; λ_{em}: 500–600 nm; two-photon excitation (c and d) (c) λ_{ex}: 840 nm; λ_{em}: 450–550 nm; (d) λ_{ex}: 840 nm; λ_{em}: 500–600 nm. Two-photon excitation increased the fluorescence of TPA-BI in TAG/LDs, while BODIPY 493/503 showed a significant signal loss.[112] (B–D) ABCXF (2 μM) stained deep tissue imaging of lipid drops in guinea pig liver tissue fed with high-fat (scale bar: 20 μm). (B) One-photon ($\lambda_{ex}=488$ nm) images; (C) two-photon ($\lambda_{ex}=850$ nm) images; (D) reconstructed 3D two-photon fluorescent images. The spherical LDs in the tissue sample were visualized along the z-axis up to a depth of 42 μm, and a high-resolution 3D two-photon fluorescent image was effectively constructed with ABCXF under two-photon excitation, whereas one-photon fluorescence imaging only showed a narrow penetration depth (<30 μm).[116]

decorating the nonaromatic rotor of trifluoromethyl group (CF_3) on the π-conjugated core, another two-photon AIE-compound, ABCXF ($C_{21}H_{16}F_6N_2$) has been developed to show better tissue penetration during lipid imaging in guinea pig liver tissue (Fig. 4B–D).[116] Compared to the one-photon fluorescence imaging (λ_{ex}: 488 nm), under excitation of 850 nm, this probe signaled better at 500–600 nm (Fig. 4A). It has been assumed that due to the intramolecular charge transfer effects, solvatochromism properties of ABCXF have increased fluorescence intensity with the increased solvent polarity that might favor this two-photon AIE probe in the biological systems.

In 2018, a new family of two-photon donor–acceptor AIEgens has been developed by Niu et al.[113] by utilizing naphthalene as the core and termed as NAP AIEgens. This family of two-photon probes includes NAP-Ph ($C_{28}H_{25}N_2O$), NAP-Br ($C_{22}H_{19}BrN_2O$), NAP-CF$_3$ ($C_{24}H_{18}F_6N_2O$), and NAP-Py ($C_{27}H_{23}N_3O$) and showed successful LDs staining performance in HeLa cells at an ultra-low concentration (50–100 nM) (Fig. 5A). Due to the hydrophobic nature of lipid drops, more hydrophobic organic dyes show more specificity for lipid droplets staining. The calculated partition coefficient (*P*) (*C* log *P*) values of NAP-Ph, NAP-Br, NAP-CF$_3$, and NAP-Py are 6.681, 5.656, 6.559 and 5.184, respectively, which is higher than the traditional lipid specific dyes Nile Red (4.618) and BODIPY 493/503 (5.028) (Fig. 5B). Since the partition coefficient (*P*) describes the dissolving ability of a compound in an immiscible biphasic system of lipid (fats, oils, organic solvents) and water, the higher *C* log *P* values of the NAP AIEgens indicate more hydrophobicity than Nile Red and BODIPY 493/503 and potentially one of the reasons of better performance at ultra-low concentration. Higher dye concentration could have detrimental effects on normal physiological activities of the cells; hence the practical applications of these two-photon dyes at lower concentration could be useful and might serve better for lipid research than other AIEgens based on the purpose of the studies.

5.3 AIE-based fluorophore to visualize lipid drops-lysosome interplay

The unique labeling behaviors and distinguished emissions properties of certain AIE probes have created new possibilities to visualize in vivo lipophagic triacylglycerol catabolism. As an integral part of cellular metabolism, the interaction between cytoplasmic LDs and lysosome can be associated with several inflammations and metabolic disorders.[119,120] Recently, a DSPE-PEG$_{2000}$ encapsulated AIE-based fluorophore TPA-BTTDO ($C_{54}H_{38}N_2O_2S_2$) has been synthesized, which shows dual emission with red and cyan fluorescence in the lysosome and LDs, respectively.[121] The molecular structure of TPA-BTTDO encompasses an electron-withdrawing thieno[3,2-*b*]thiophene *S*,*S*-dioxide (TTDO) core and electron-donating triphenylamine (TPA) groups. This enables the molecule for the twisted intramolecular charge transfer (TICT) between TTPO and TPA, particularly in a high polar environment. Due to the poor water solubility, to increase the dispersibility of water solvent in the biological system, the AIEgen has been encapsulated in DSPE-PEG$_{2000}$.

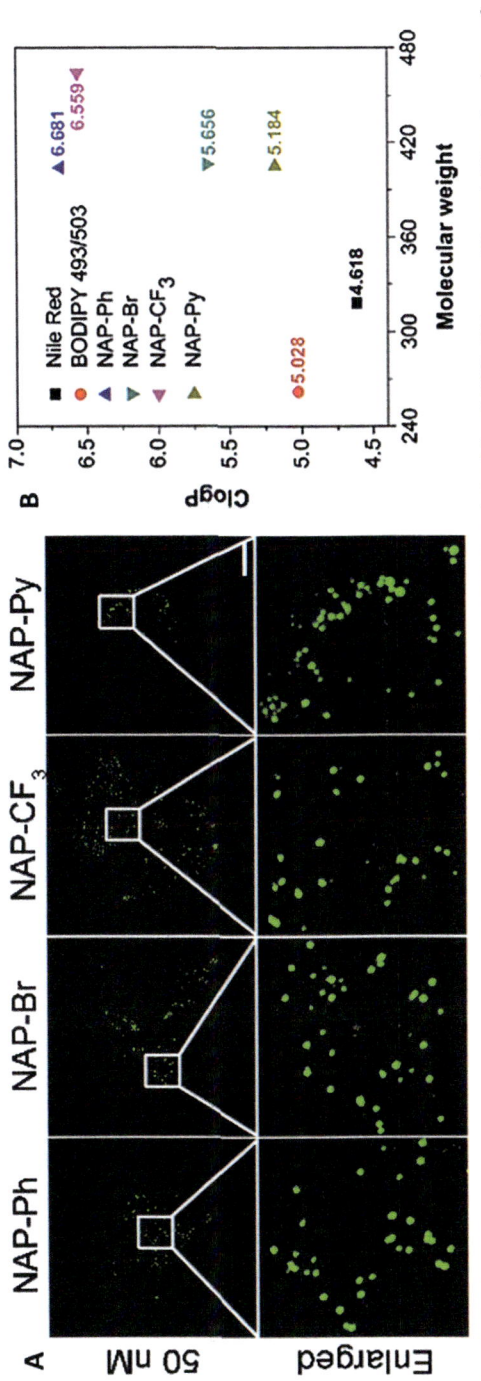

Fig. 5 Confocal laser scanning microscopy of lipid drops with ultra-low concentration (50 nM) of NAP-AIEgens (A) and C log P values of NAP-AIEgens (B). (A) Images of HeLa cells stained with NAP-Ph, NAP-Br, NAP-CF3, NAP-Py. Scale bar: 20 μm. (B) C log P of Nile Red, BODIPY 493/503, NAP-Ph, NAP-Br, NAP-CF$_3$, and NAP-Py. Higher values are indicator of better lipid droplet staining performance.[113]

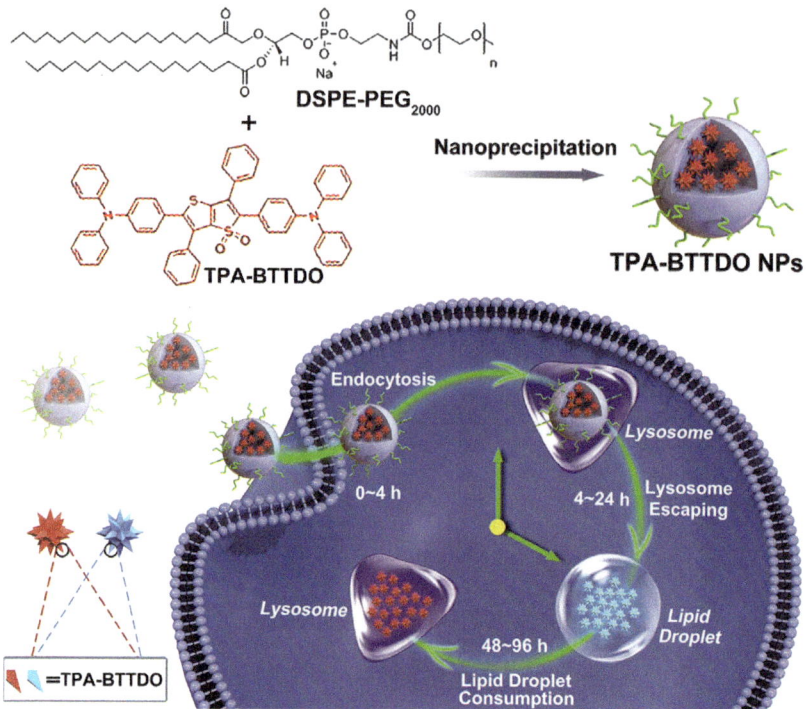

Fig. 6 Systematic diagram of the internalization and consumption processes of TPA-BTTDO NPs in cells.[121]

Following the internalization through the cell membrane via endocytosis, during the localization in the lysosome, the probe shows high polarity dependant red emission (614 nm). After enzymatic degradation of DSPE-PEG$_{2000}$ matrix, the exposure of hydrophobic probe TPA-BTTDO to the less polar hydrophobic LDs results in cyan emission (498 nm). These properties enable us to conceptualize LD components degradation by lysosomal enzymes (Fig. 6), which could hold significant prospects for the investigation of LDs–lysosome related metabolic disorders.

5.4 Theranostics approaches with lipid-specific AIE probes

Fluorescence-guided theranostics approaches combine both therapeutic and diagnostic processes within one platform, therefore conciliates prominent opportunities for early disease diagnosis and visible drug delivery in noninvasive, and fast responsive manners. It also provides valuable insights into the biological structures and multiplexed processes. However, currently available conventional fluorophores for these purposes generally significantly suffer from ACQ phenomena, where their emission quenches upon aggregation due to the intermolecular π-π stacking and other nonradiative pathways. To minimize these limitations, AIE molecules have been a research focus and several AIEgens have been recently introduced concerning photodynamic therapy (PDT) and disease treatments that have been discussed in this section.

PDT is a noninvasive and reliable alternative cancer therapeutic technique that uses photosensitizers to destroy the targeted tissues. Compared to the non-transformed cells in the majority of cancer cells, biochemical machinery required for effectually detoxifying the H_2O_2 are low and thus the amount is higher in tumor cells.[122] Also in recent years, several studies have demonstrated an increasing amount of intracellular lipid accumulation in different neoplastic processes.[22] Given these facts, recent fabrication of LD-targeted H_2O_2-activatable fluorophore, TPECNPB offers the great opportunity to treat cancer through singlet oxygen generation.[123] This unique AIE-activated photosensitizer exhibits a massive Stokes shift of 175 nm, showed remarkable photostability, and excellent selectivity toward H_2O_2 with aggregation quenching free effects. Therefore, fluorescence-guided PDTs have been predicted to be more promising in cancer cells with TPECNPB (Fig. 7).

Moreover, fluorophores with red/near-infrared (NIR) region emission have been considered promising for biological applications as they could reduce the photodamage to the biological samples and avoid autofluorescence of the physiological environments. At the same time, they are featured to improve optical absorption interference and reduce the light scattering that results in bright fluorescence. AIE characteristics also allow these probes to work perfectly in aggregated states or at high concentrations with a high photobleaching threshold.[124,125] Generally, strategies of the red/NIR fluorophores synthesis are incredibly tedious, time, cost, and energy-consuming that require strong connections among electron-donating and -accepting units through π-bridge(s), expansion of π-conjugation, and combination of these schemes.[126,127] Recently, a one-pot synthetic strategy to LD-specific red/NIR AIEgens, TPMN ($C_{28}H_{19}N_3$), TTMN ($C_{26}H_{17}N_3S$), MeTTMN ($C_{28}H_{21}N_3S$), and MeOTTMN ($C_{28}H_{21}N_3O_2S$) with simple structures has been reported by Wang et al.[127] Upon aggregation, the emission of these probes range from 648 to 719 nm. In contrast to the BODIPY 493/503 Green, TTMN and MeTTMN showed 65% more photostability after constant irradiation for 15 min with 40 scans, while TPMN and MeOTTMN were 49% and 54% more photostable than BODIPY 493/503 Green. Additionally, these AIEgens have shown bright emission even with a

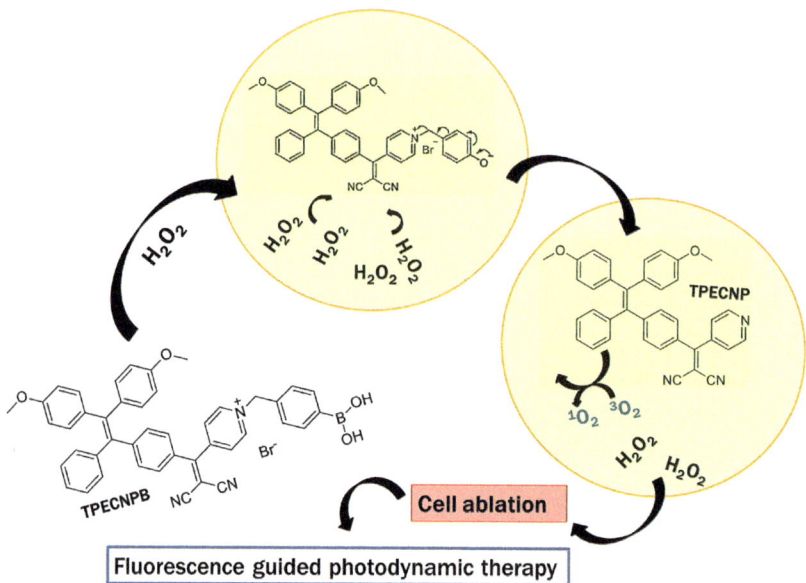

Fig. 7 Working mechanism of TPECNPB. TPECNPB, with an amphiphilic architecture and a positively charged pyridinium pendent target the amphipathic and negatively charged lipid drops through electrostatic interactions. Boronate ester in pyridinium pendant of TPECNPB molecule works as the H_2O_2 activatable site. After localization in lipid drops, high levels of H_2O_2 in hypoxia cancer cells liberate the hydrophobic AIEgen TPECNP that emits red emission due to aggregation. As a consequence, of fluorescence-guided light irradiation, cancer cell apoptosis can be induced by photo-generated singlet oxygen ($^1O_2$) by AIE-active TPECNP.[123]

very lower concentration of 200×10^{-9} M and could kill cancer cells by generating higher amounts of reactive oxygen species (ROS). Therefore, these NIR-AIEgens could be ideal as LD-specific photosensitizers for photodynamic ablation of cancer cells at very low concentration.

5.5 Lipid-specific AIEgens system with wide emission tunability

Recently, designing AIE materials with red/NIR emissive properties has become of significant interest to the AIE researchers. Being one of the three primary colors (red, yellow, blue), red is the crucial component and has been utilized in a wide range of optoelectronic devices. Due to the higher penetrability, lower background noise and minimal cellular damage, fluorescence materials emitted at red/NIR regions are more preferential for biological studies. AIE materials with red/NIR emissive characteristics are also privileged for inorganic quantum dots due to lower toxicity, higher emission and photobleaching resistance. However, despite significant efforts, due to the tedious synthesis process and complex structures few lipid-specific probes with red/NIR emissive properties have been reported up-to-date. Recently, by incorporating phenol, ethene-1,1,2-triyltribenzene, and diphenylamine in the ACQ compound, Zhang et al.[128] has developed an AIE-system by stepwise enhancing the donor–acceptor interactions. This strategy covers almost all regions of visible light and pursues maximum emission in the NIR region (Fig. 8A and B). In the study, structures with intramolecular motions, such as oxydibenzene, triphenylamine (TPA), and tetraphenylethene (TPE) were donor, the carbon–carbon double bond acted as the π bridge, and cyanogroup, trifluoromethyl, as well as the benzene ring, attached to the nitro group acted as the acceptor. These AIEgens showed very lower cytotoxicity and superior specificity to the LD. AIEgen, B3 showed high sensitivity to trioctanoin, an increased triglyceride found in hyperlipidemia patients. Therefore, this AIEgen enables us to differentiate the hyperlipidemia patient's and normal people's blood (Fig. 8C and D).

5.6 Biocompatible AIEgen from natural resources

Fluorogenic compounds derived from natural resources could be environmental-friendly, biodegradable and cost-effective, which might increase their opportunities for applications. Additionally, exploration of water-soluble AIEgens can hold fundamental advantages for biological studies. A recent study has reported the extraction of a natural isoquinoline alkaloid, berberine chloride (BBR chloride) from Chinese herbs that showed unconventional rotor-free AIEgen characteristic with water-soluble properties.[129]

In dilute water solution, BBR chloride shows weak emission, which possibly occurs due to the active intramolecular vibration and the TICT effect that ways to the non-radiative decay. Upon aggregation, the pathway of the non-radiative decay in this AIEgen is intimidated and causes bright fluorescence. With the increasing solvent polarity, there is almost no change in the absorption of this fluorophore, but the fluorescence intensity decreases significantly with a red shifting of the emission maximum from 525 nm to 550 nm. BBR chloride also shows sensitivity to viscosity and temperature. Upon exposure to the more viscous environment and lower temperature,

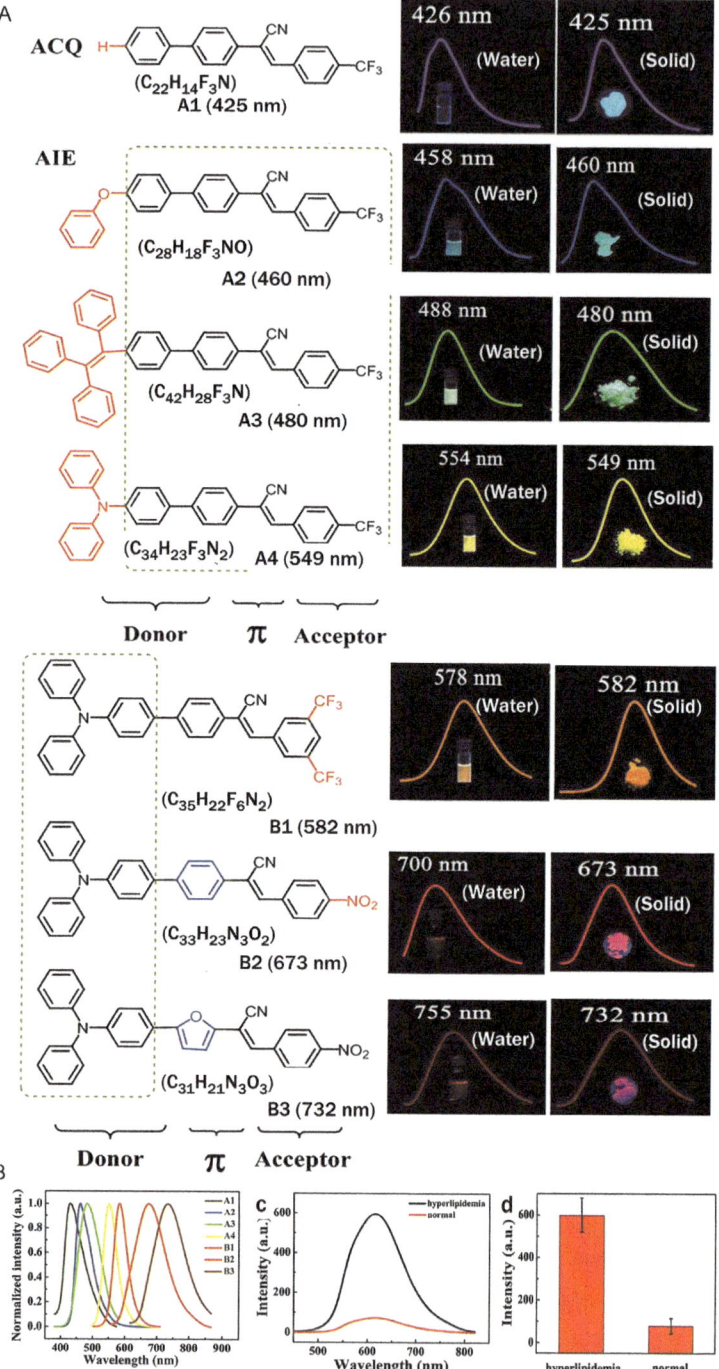

Fig. 8 See figure legend on opposite page.

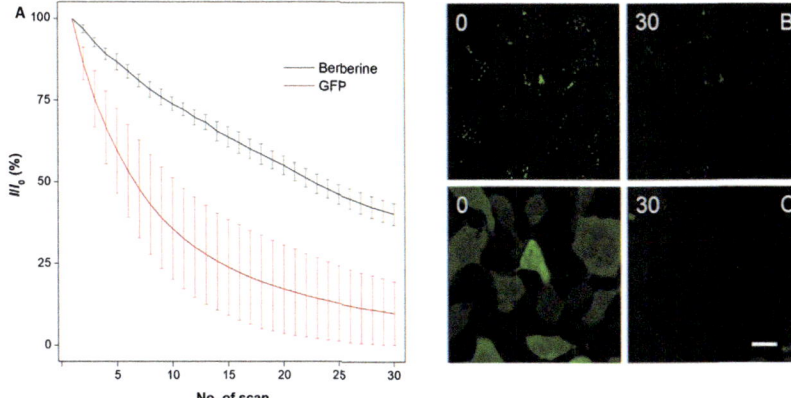

Fig. 9 Photostability of berberine chloride (BBR chloride) and green fluorescent protein (GFP). (A) Continuous scanning of BBR Chloride and GFP at 488 nm (2.3 μW). I_0: initial PL intensity; I: PL intensity of the corresponding sample after a designated no. of a scan. (B) HeLa cells stained with 10 μM BBR Chloride and (C) 786-O cells with GFP gene before and after 30 scans of light irradiation. λ_{ex}: 488 nm; scale bar: 20 μm. Error bars ± relative standard deviations, $n=6$.[129]

the fluorescence intensity of BBR chloride shows a sharp increase, which signifies the influence of the intramolecular vibration in this AIEgen as well. This biocompatible fluorophore can successfully stain lipid drops in different cells, such as HeLa, A549 and MCF-10A cells. In comparison to the green fluorescent protein (GFP), after 20 scans, BBR chloride retained 30% more fluorescence intensity, proving the photostable character of this probe is over the traditional fluorophore, GFP (Fig. 9). These characters make this naturally obtained LDs specific AIEgen a promising candidate for LD imaging and associated disease diagnosis.

Fig. 8 Chemical structures of the tunable AIE system from an ACQ compound (A1) to AIEgens (A1, A2, A3, A4, B1, B2 and B3) and their emissions. (A) Tuning AIEgens (A1, A2, A3, A4, B1, B2 and B3) from compound (A1); respective photographs of in water and solid state were taken under 365 nm UV light. Installing a phenol rotor to the ACQ compound, A1 constructs the AIE-active compound A2, while the introduction of diphenylamine and ethene-1,1,2-triyltribenzene rotors construct AIEgens A3 and A4, respectively. Subsequently, the electron-withdrawing group of A4 was modified to get compounds B1, B2 and B3 and the emission reached in NIR region. (B) Normalized PL spectra of A1 (λ_{em}: 425 nm), A2 (λ_{em}: 460 nm), A3 (λ_{em}: 480 nm), A4 (λ_{em}: 549 nm), B1 (λ_{em}: 582 nm), B2 (λ_{em}: 673 nm) and B3 (λ_{em}: 732 nm) in the solid state. (C) Emission spectra of B3 (1.0×10^{-6} M, λ_{ex}: 423 nm) in the serum of hyperlipidemia and normal people. (D) Bar diagram of the fluorescence intensities in the serum of hyperlipidemia and normal people ($n=6$).[128]

5.7 Lipid specific AIEgens in algae research

Microalgae have diverse benefits and a high potential for biofuel production. Unlike the land-based plants, green microalgae could be utilized more effectively for biofuel and health beneficiary food supplementation. Over the past few years, numbers of studies have been reported to illustrate different lipid-specific AIE probes in medical research for disease diagnosis and photodynamic therapy in animal cell models. However, the reports on AIE molecules introduced to microalgae are still very few and a significant area to explore. Different microalgae have their unique cell types and autofluorescence characteristics, so the requirement of appropriate selection of AIEgens and their acquisition techniques are crucial for studying lipid drops in microalgae.

The study in an algal model of *Nannochloropsis* sp. with AIE probe TPE-AmAl, revealed greenish-blue emission upon aggregation of the dye in lipid drops (Fig. 10A and B). Among different concentration of DMSO, higher contents showed shorter diffusion times to assist the probe's entry into the algae through the cell wall.[110] In another research, a microalga, *Euglena gracilis* has been induced for more health beneficiary fatty acids synthesis.[130] In that study, a novel strategy for easy and rapid detection of lipid drops with AIEgen, DPAS has been used to identify maximum lipid synthesis condition among different treatments (Fig. 10C–L).

Considering the photostability and sample preparation techniques, luminogen DPAS has surpassed the lipid drops staining performances of the commercial fluorophore, BODIPY. This study also has signified an easy technique for the lipid study in this alga. Application of a much higher concentration of the probe of 100 μM showed no deleterious impact on the growth or behavior of *Euglena* cells or to the fluorescence intensity. DPAS also showed a very low background signal and was found more sensitive than BODIPY for semiquantitative in vivo fluorescence measurements. Co-staining with the BODIPY and chlorophyll also indicated its effectiveness for multicolor imaging with red and green fluorophores. Additionally, sample preparation steps with DPAS is time-efficient and less stressful for the living organism as it involved only a fewer wash free steps than that of the thorough washing requirements of BODIPY dye. This might facilitate the study of other physiological factors within the cell in a natural state. A brief summary of the lipid specific AIE probes is given in Table 2.

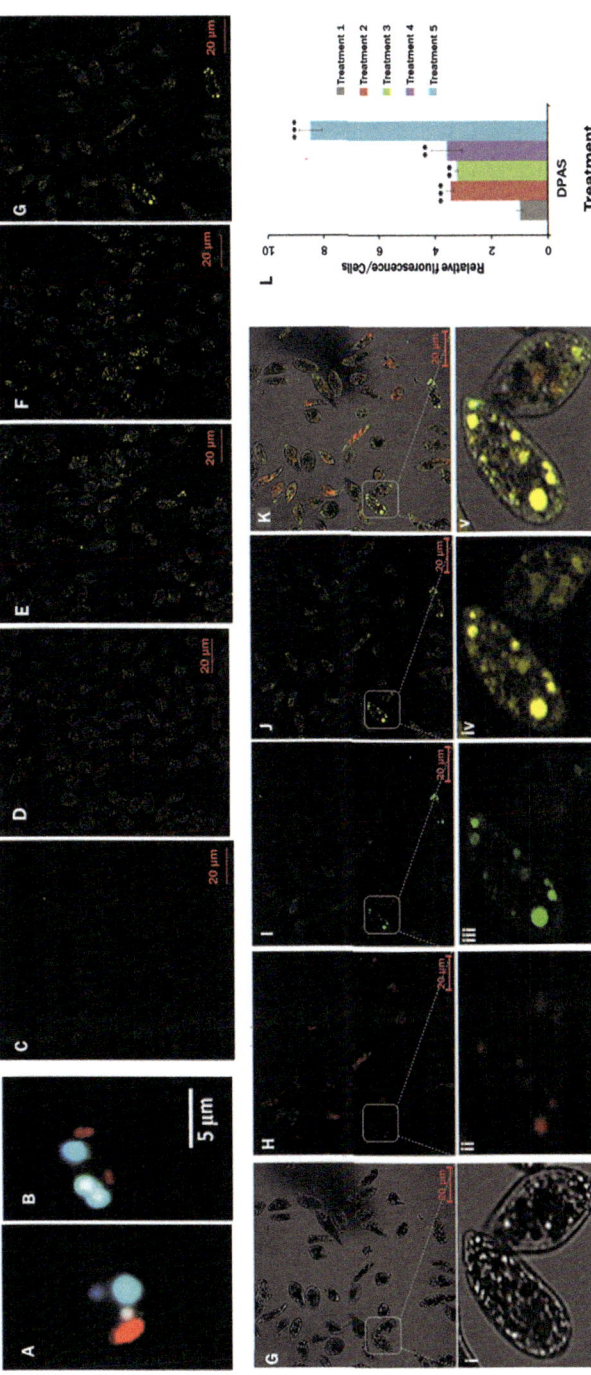

Fig. 10 Study of lipid drops in microalgae with AIEgens. (A and B) Stained *Nannochloropsis* sp. incubated with 2.5 μM TPE-AmAI for 10 min at 40 °C in (A) 10 vol% and (B) 20 vol% of DMSO. λ_{ex}: 330–385 nm. The blue and red emissions are from the LDs and chloroplast, respectively.[110] (C–L) *Euglena gracilis* cells stained with DPAS (10 μM). (C) Treatment 1: modified Cramer-Myers medium (MCM); (D) Treatment 2: MCM, (−) N_2; (E) Treatment 3: MCM, (−) N_2, (−) Ca^{2+}; (F) Treatment 4: MCM, (−) N_2, (−) Ca^{2+}, (+) glucose (24 h light); (G) Treatment 5: MCM, (−) N_2, (−) Ca^{2+}, (+) Glucose (24 h dark) conditions. (G–K) Bright-field, fluorescence and merged images lipid-induced cells. Bright-field image: G; Fluorescence images—Chlorophyll: H, BODIPY: I, DPAS: J; Merged images: K, and the enlarged regions of respective images (i, ii, iii, iv, and v). (L) Relative fluorescence intensity/cell of DPAS for different treatments. Values are relative to the control condition (Treatment 1: modified Cramer–Myers medium (MCM)). Averages shown as mean ± SE; $*P < 0.05$; $**P < 0.01$.[130]

Table 2 Properties of some recently synthesized LDs specific AIE fluorescent probes.

Probes	Chemical structure	$\lambda_{ex}/\lambda_{em}$ (nm)	Solubility	Notes
FAS[109]	Chemical formula: $C_{20}H_{14}N_2O$ Calculated MW: 298.33	λ_{ex}: 400–440; λ_{em}: 595	DMSO	• Lesser background noises • Higher photostability • Biocompatible and almost no cytotoxicity
DPAS[109]	Chemical formula: $C_{20}H_{16}N_2O$ Calculated MW: 300.35	λ_{ex}: 330–385; λ_{em}: 565	DMSO	• Much lower background noises • No noticeable cytotoxicity • Photostable and biocompatible
TPE-AmAI[110]	Chemical formula: $C_{31}H_{30}N_2O$ Calculated MW: 446.2358	λ_{ex}: 330–385; λ_{em}: ≥420	• DMSO • THF • $CHCl_3$ • CH_2Cl_2 • DMF	• Sensitive to the environment polarity • Higher photostability • Lower background noise • Biocompatible and almost no cytotoxicity

TPE-AC[111]	Calculated MW: 508.2627	λ_{ex}: 455 in THF; 510–550 in DMSO. λ_{em}: ≥705	• DMSO • THF	• Bright near-infrared emission • Higher photostability • Biocompatible; very low cytotoxicity
TPA-BI[112]	Chemical formula: $C_{32}H_{29}N_3O$ Calculated MW: 471.2311	λ_{ex}: 400–440 and 840; λ_{em}: 450–550	DMSO and other polar solvents	• Competent for one-photon and two-photon imaging • Reduced photobleaching • Deeper tissue penetration • Photophysical properties are related to solvent polarity
ABCXF[116]	Chemical formula: $C_{21}H_{16}F_6N_2$ Calculated MW: 410.1218	λ_{ex}: 488 and 850; λ_{em}: 500–600	DMSO, DCM, acetonitrile, acetone, EtOH, DMF MeOH, THF, toluene, chloroform, ethyl ether	• Competent for one-photon and two-photon imaging • Deeper tissue penetration up to a depth of 42 μm in two-photon imaging • Reduced photobleaching • Increased fluorescence intensity with increased polarity • More suitable in the biological system

Continued

Table 2 Properties of some recently synthesized LDs specific AIE fluorescent probes.—cont'd

Probes	Chemical structure	$\lambda_{ex}/\lambda_{em}$ (nm)	Solubility	Notes
TPA-BTTDO[121]	Chemical formula: $C_{54}H_{38}N_2O_2S_2$ Calculated MW: 810.2375	λ_{ex}: 405; λ_{em}: 450–725	THF, CH_2Cl_2, DMF and other organic solvents	• Sensitive to the environment polarity • Cyan emission (498 nm) in lipid droplet; red emission (614 nm) in the lysosome • Photostable and biocompatible
TPECNPB[123]	Chemical formula: $C_{44}H_{35}BN_3O_4^+$ Calculated MW: 680.2715	λ_{ex}: 450; λ_{em}: 625	THF	• Sensitive to H_2O_2, enable cancer cell ablation efficiently through ROS generation • Photostable, support long-term visualization of the photodynamic therapy efficiently in vitro and in vivo
Berberine chloride[129]	·2H₂O Chemical formula: $C_{20}H_{18}ClNO_4$ Calculated MW: 371.81	λ_{ex}: 488; λ_{em}: 500–580	H_2O	• Water soluble, more suitable for biological studies • Photostable and biocompatible • Sensitive to the viscosity and temperature

DMA-POABP[131]	![structure] Chemical formula: $C_{16}H_{16}BF_2N_3O$ Calculated MW: 316.1354	λ_{ex}: 330–385; λ_{em}: 448–548	Hexane, toluene, DCM, 1,4-dioxane, chloroform, THF, ACN, DMF, EtOH, MeOH, IPA, DMSO	• Organoboron isomer • In HeLa cells showed better resolution and higher contrast in comparison with Nile Red • Bright-blue fluorescence • Biocompatible, no noticeable repressive effect on cell • More photostable than Nile Red • Strong emission in a viscous medium • In solution, exhibited bluer but weaker emission than POABP-DMA
POABP-DMA[131]	![structure] Chemical formula: $C_{16}H_{16}BF_2N_3O$ Calculated MW: 315.1354	λ_{ex}: 330–385; λ_{em}: 448–548	Hexane, toluene, DCM, 1,4-dioxane, chloroform, THF, ACN, DMF, EtOH, MeOH, IPA, DMSO	• Organoboron isomer • In HeLa cells showed better resolution and higher contrast in comparison with Nile Red • Bright-blue fluorescence • Biocompatible, no noticeable repressive effect on cell • More photostable than Nile Red • Strong emission in viscous medium • In solution, exhibited less bluer but stronger emission than DMA-POABP

Continued

Table 2 Properties of some recently synthesized LDs specific AIE fluorescent probes.—cont'd

Probes	Chemical structure	$\lambda_{ex}/\lambda_{em}$ (nm)	Solubility	Notes
AP-DEA[132]	Chemical formula: $C_{18}H_{21}N_3O$ Calculated MW: 295.1685	λ_{ex}: 405; λ_{em}: 448–548	THF, DMSO	• High colocalization with Nile Red in HeLa cells • Superior photostability • Can generate reactive oxygen species very fast and effectively, therefore much suitable for photodynamic therapy • No noticeable repressive effect on HeLa cell growth
NAP-AIEgen: NAP-Ph, NAP-Br, NAP-CF3, NAP-Py[113]				
NAP-Ph	Chemical formula: $C_{28}H_{25}N_2O$ Calculated MW: 405.1961			

NAP-BR	Chemical formula: $C_{22}H_{19}BrN_2O$ Calculated MW: 406.0691	λ_{ex}: 405 and 860; λ_{em}: 480–560	THF, CH_3CN, toluene, diethyl ether, acetone, DMSO	• Competent for one-photon and two-photon imaging • Deeper tissue penetration up to a depth of 70 μm in two-photon imaging has been recorded for NAP-CF$_3$ • Reduced photobleaching • Could be used at an ultra-low concentration (50–100 nM)
NAP-CF$_3$	Chemical formula: $C_{24}H_{18}F_6N_2O$ Calculated MW: 464.1323			
NAP-Py	Chemical formula: $C_{27}H_{23}N_3O$ Calculated MW: 405.1841			

Continued

Table 2 Properties of some recently synthesized LDs specific AIE fluorescent probes.—cont'd

Probes	Chemical structure	$\lambda_{ex}/\lambda_{em}$ (nm)	Solubility	Notes
One-pot synthesized AIEgens: TPMN, TTMN, MeTTMN, and MeOTTMN[127]				
TPMN	Chemical formula: $C_{28}H_{19}N_3$ Calculated MW: 397.1579	λ_{ex}: 441; λ_{em}: 637	Toluene, ethyl acetate, THF, chloroform, DMSO, acetone, DCM, acetonitrile, MeOH	• Red/NIR AIE luminogens • Simple structures • Photostable and bright emission • Suitable for photodynamic therapy • Low toxicity in dark
TTMN	Chemical formula: $C_{26}H_{17}N_3S$ Calculated MW: 403.1143	λ_{ex}: 483; λ_{em}: 672		
MeTTMN	Chemical formula: $C_{28}H_{21}N_3S$ Calculated MW: 431.5570	λ_{ex}: 492; λ_{em}: 681		

MeOTTMN λ_{ex}: 499; λ_{em}: 701

Chemical formula:
$C_{28}H_{21}N_3O_2S$
Calculated MW:
463.1354

AIE system of A2, A3, A4, B1, B2 and B3 with wide tunable emissions[128]

A2 λ_{ex}: 405; λ_{em}: 420–480

Chemical formula:
$C_{28}H_{18}F_3NO$
Calculated MW:
442.1413

Continued

Table 2 Properties of some recently synthesized LDs specific AIE fluorescent probes.—cont'd

Probes	Chemical structure	$\lambda_{ex}/\lambda_{em}$ (nm)	Solubility	Notes
A3	Chemical formula: $C_{42}H_{28}F_3N$ Calculated MW: 604.2247	λ_{ex}: 458; λ_{em}: 580–650		• Tunable AIEgens with wide visible spectra (400–780 nm) • High photostability and bright emission • Low toxicity • B3 achieved far-red emission • B3 is sensitive to trioctanoin, therefore can distinguish hyperlipidemia patient blood
A4	Chemical formula: $C_{34}H_{23}F_3N_2$ Calculated MW: 517.1886	λ_{ex}: 514; λ_{em}: 520–570	DMSO	

B1

λ_{ex}: 514; λ_{em}: 540–620

Chemical formula:
$C_{35}H_{22}F_6N_2$
Calculated MW:
585.1760

B2

λ_{ex}: 514; λ_{em}: 590–680

Chemical formula:
$C_{33}H_{23}N_3O_2$
Calculated MW:
494.1863

B3

λ_{ex}: 488; λ_{em}: 600–750

Chemical formula:
$C_{31}H_{21}N_3O_3$
Calculated MW:
484.1656

6. Conclusions

The emergence of nanoprobes with AIE attributes in recent years is promising in remunerative characteristics with defined photosensitizer properties. With the apprehension of the increasing necessities, AIE studies are currently expanding their spheres to facilitate inimitable prospects in biomedical domains. By utilizing the easy synthetic techniques and tuning the AIE properties, highly selective, stable and facile nano-assemblies could be fabricated for bioimaging and therapeutic functions for greater human benefits.

7. Future remarks

Despite the visible progress for the animal cells, reports on the applications of AIEgens to study lipid drops in algae are limited. The study may vary based on cell wall structure that might have different permeability to different AIE molecules. Some algae cells have protein and glycan based cell membrane, and others have silica and organic components based cell wall. Therefore, the penetration and internalization of the AIE fluorophores might be algal species specific and require in-depth exploration to incorporate this nanotechnology and make further advancement in algal lipidomics. This will not only play an essential role to bring two emerging fields of nanoscience and algae research together, but also will establish a new platform in algal research that could be outreached to different geographic regions.

Acknowledgments

AHM Mohsinul Reza is grateful for the financial support of Australian Government Research Training Program Scholarship (AGRTPS) (International) for his PhD study at Flinders University, Australia.

References

1. Cinti S, Giordano A. The adipose organ. In: Longhi S, et al., eds. *The First Outstanding 50 Years of "Università Politecnica Delle Marche"*. Cham: Springer; 2020.
2. Zhang C, Yang L, Ding Y, et al. Bacterial lipid droplets bind to DNA via an intermediary protein that enhances survival under stress. *Nat Commun.* 2017;8:15979.
3. Thiele C, Spandl J. Cell biology of lipid droplets. *Curr Opin Cell Biol.* 2008;20:378–385.
4. Olzmann JA, Carvalho P. Dynamics and functions of lipid droplets. *Nat Rev Mol Cell Biol.* 2019;20:137–155.

5. Robenek MJ, Severs NJ, Schlattmann K, et al. Lipids partition caveolin-1 from ER membranes into lipid droplets: updating the model of lipid droplet biogenesis. *FASEB J*. 2004;18:866–868.
6. Wan HC, Melo RC, Jin Z, Dvorak AM, Weller PF. Roles and origins of leukocyte lipid bodies: proteomic and ultrastructural studies. *FASEB J*. 2007;21:167–178.
7. Guo Y, Cordes KR, Farese Jr RV, Walther TC. Lipid droplets at a glance. *J Cell Sci*. 2009;122:749–752.
8. Gao Q, Goodman JM. The lipid droplet—a well-connected organelle. *Front Cell Dev Biol*. 2015;3:49.
9. Ohsaki Y, Suzuki M, Fujimoto T. Open questions in lipid droplet biology. *Chem Biol*. 2014;21:86–96.
10. Bulger RE, Strum JM. The machinery of the cytoplasm. In: *The Functioning Cytoplasm*. Boston, MA: Springer; 1974. Plenum Press; New York.
11. Farese Jr RV, Walther TC. Lipid droplets finally get a little R-E-S-P-E-C-T. *Cell*. 2009;139:855–860.
12. Yang L, Ding Y, Chen Y, et al. The proteomics of lipid droplets: structure, dynamics, and functions of the organelle conserved from bacteria to humans. *J Lipid Res*. 2012;53:1245–1253.
13. Ruocco N, Albarano L, Esposito R, Zupo V, Costantini M, Ianora A. Multiple roles of diatom-derived oxylipins within marine environments and their potential biotechnological applications. *Mar Drugs*. 2020;18:342.
14. Reza AHMM, Tavakoli J, Zhou Y, Qin J, Tang Y. Synthetic fluorescent probes to apprehend calcium signalling in lipid droplet accumulation in microalgae-an updated review. *Sci China Chem*. 2020;63:308–324.
15. Goncalves EC, Wilkie AC, Kirst M, Rathinasabapathi B. Metabolic regulation of triacylglycerol accumulation in the green algae: identification of potential targets for engineering to improve oil yield. *Plant Biotechnol J*. 2016;14:1649–1660.
16. Roingeard P, Melo RC. Lipid droplet hijacking by intracellular pathogens. *Cell Microbiol*. 2017;19:e12688.
17. Johnson MR, Stephenson RA, Ghaemmaghami S, Welte MA. Developmentally regulated H2Av buffering via dynamic sequestration to lipid droplets in *Drosophila* embryos. *eLife*. 2018;7:e36021.
18. Goldberg IJ, Reue K, Abumrad NA, et al. Deciphering the role of lipid droplets in cardiovascular disease: a report from the 2017 National Heart, Lung, and Blood Institute Workshop. *Circulation*. 2018;138:305–315.
19. Akinci B, Sahinoz M, Oral E. Lipodystrophy syndromes: presentation and treatment. In: Feingold KR, et al., eds. *Endotext*. MDText.com, Inc; 2018.
20. Missaglia S, Coleman RA, Mordente A, Tavian D. Neutral lipid storage diseases as cellular model to study lipid droplet function. *Cell*. 2019;8:187.
21. Fanning S, Haque A, Imberdis T, et al. Lipidomic analysis of α-synuclein neurotoxicity identifies stearoyl CoA desaturase as a target for Parkinson treatment. *Mol Cell*. 2019;73:1001–1014.e8.
22. Cruz A, Barreto EA, Fazolini N, Viola J, Bozza PT. Lipid droplets: platforms with multiple functions in cancer hallmarks. *Cell Death Dis*. 2020;11:105.
23. Luterbacher JS, Martin AD, Dumesic JA. Targeted chemical upgrading of lignocellulosic biomass to platform molecules. *Green Chem*. 2014;16:4816–4838.
24. Singh A, Olsen SI. A critical review of biochemical conversion, sustainability and life cycle assessment of algal biofuels. *Appl Energy*. 2011;88:3548–3555.
25. Phukan MM, Chutia RS, Konwar BK, Kataki R. Microalgae *Chlorella* as a potential bio-energy feedstock. *Appl Energy*. 2011;8:3307–3312.
26. Jones CS, Mayfieldt SP. Algae biofuels: versatility for the future of bioenergy. *Curr Opin Biotechnol*. 2012;23:346–351.

27. Ji RY, Ding Y, Shi TQ, et al. Metabolic engineering of yeast for the production of 3-hydroxypropionic acid. *Front Microbiol.* 2018;9:2185.
28. Du ZY, Alvaro J, Hyden B, et al. Enhancing oil production and harvest by combining the marine alga *Nannochloropsis oceanica* and the oleaginous fungus *Mortierella elongata*. *Biotechnol Biofuels.* 2018;11:174.
29. Hwangbo M, Chu KH. Recent advances in production and extraction of bacterial lipids for biofuel production. *Sci Total Environ.* 2020;734:139420.
30. Khan MI, Shin JH, Kim JD. The promising future of microalgae: current status, challenges, and optimization of a sustainable and renewable industry for biofuels, feed, and other products. *Microb Cell Factories.* 2018;17:36.
31. Patnaik R, Mallick N. Utilization of *Scenedesmus obliquus* biomass as feedstock for biodiesel and other industrially important co-products: an integrated paradigm for microalgal biorefinery. *Algal Res.* 2015;12:328–336.
32. Arroussi HEL, Benhima R, Iman B, Mernissi NE, Wahby I. Improvement of the potential of *Dunaliella tertiolecta* as a source of biodiesel by auxin treatment coupled to salt stress. *Renew Energy.* 2015;77:15–19.
33. Farrokheh A, Tahvildari K, Nozari M. Biodiesel production from the *Chlorella vulgaris* and *Spirulina platensis* microalgae by electrolysis using $CaO/KOH-Fe_3O_4$ and $KF/KOH-Fe_3O_4$ as magnetic nanocatalysts. *Biomass Convers Biorefin.* 2020. https://doi.org/10.1007/s13399-020-00688-z.
34. Cabanelas IT, van der Zwart M, Kleingris DM, Wijffels RH, Barbosa MJ. Sorting cells of the microalga *Chlorococcum littorale* with increased triacylglycerol productivity. *Biotechnol Biofuels.* 2016;9:183.
35. Chiu SY, Kao CY, Tsai MT, Ong SC, Chen CH, Lin CS. Lipid accumulation and CO_2 utilization of *Nannochloropsis oculata* in response to CO_2 aeration. *Bioresour Technol.* 2009;100:833–838.
36. Hosseini A, Jazini M, Mahdieh M, Karimi K. Efficient superantioxidant and biofuel production from microalga *Haematococcus pluvialis* via a biorefinery approach. *Bioresour Technol.* 2020;306:123100.
37. Dijkstra AJ. Revisiting the formation of trans isomers during partial hydrogenation of triacylglycerol oils. *Eur J Lipid Sci Technol.* 2006;108:249–264.
38. Chisti Y. Biodiesel from microalgae. *Biotechnol Adv.* 2007;25:294–306.
39. Knothe G. Improving biodiesel fuel properties by modifying fatty ester composition. *Energy Environ Sci.* 2009;2:759–766.
40. Galasso C, Gentile A, Orefice I, et al. Microalgal derivatives as potential nutraceutical and food supplements for human health: a focus on cancer prevention and interception. *Nutrients.* 2019;11:1226.
41. Sokoła-Wysoczańska E, Wysoczański T, Wagner J, et al. Polyunsaturated fatty acids and their potential therapeutic role in cardiovascular system disorders—a review. *Nutrients.* 2018;10:1561.
42. Rimm EB, Appel LJ, Chiuve SE, et al. Seafood long-chain n-3 polyunsaturated fatty acids and cardiovascular disease: a science advisory from the American Heart Association. *Circulation.* 2018;138:e35–e47.
43. Freitas R, Campos MM. Protective effects of omega-3 fatty acids in cancer-related complications. *Nutrients.* 2019;11:945.
44. Avallone R, Vitale G, Bertolotti M. Omega-3 fatty acids and neurodegenerative diseases: new evidence in clinical trials. *Int J Mol Sci.* 2019;20:4256.
45. Fussbroich D, Colas RA, Eickmeier O, et al. A combination of LCPUFA ameliorates airway inflammation in asthmatic mice by promoting pro-resolving effects and reducing adverse effects of EPA. *Mucosal Immunol.* 2020;13:481–492.
46. Balić A, Vlašić D, Žužul K, Marinović B, Bukvić Mokos Z. Omega-3 versus omega-6 polyunsaturated fatty acids in the prevention and treatment of inflammatory skin diseases. *Int J Mol Sci.* 2020;21:741.

47. Tricò D, Mengozzi A, Nesti L, et al. Circulating palmitoleic acid is an independent determinant of insulin sensitivity, beta cell function and glucose tolerance in non-diabetic individuals: a longitudinal analysis. *Diabetologia.* 2020;63:206–218.
48. Frigolet ME, Gutiérrez-Aguilar R. The role of the novel lipokine palmitoleic acid in health and disease. *Adv Nutr.* 2017;8:173S–181S.
49. Silva JR, Burger B, Kühl C, Candreva T, Dos Anjos M, Rodrigues HG. Wound healing and omega-6 fatty acids: from inflammation to repair. *Mediat Inflamm.* 2018;2018:2503950.
50. Guo EL, Katta R. Diet and hair loss: effects of nutrient deficiency and supplement use. *Dermatol Pract Concept.* 2017;7:1–10.
51. Weiser MJ, Butt CM, Mohajeri MH. Docosahexaenoic acid and cognition throughout the lifespan. *Nutrients.* 2016;8:99.
52. Sun GY, Simonyi A, Fritsche KL, et al. Docosahexaenoic acid (DHA): an essential nutrient and a nutraceutical for brain health and diseases. *Prostaglandins Leukot Essent Fat Acids.* 2018;136:3–13.
53. Gutiérrez S, Svahn SL, Johansson ME. Effects of omega-3 fatty acids on immune cells. *Int J Mol Sci.* 2019;20:5028.
54. Sztalryd C, Brasaemle DL. The perilipin family of lipid droplet proteins: gatekeepers of intracellular lipolysis. *Biochim Biophys Acta Mol Cell Biol Lipids.* 1862;2017:1221–1232.
55. Stone SJ, Levin MC, Zhou P, Han J, Walther TC, Farese Jr RV. The endoplasmic reticulum enzyme DGAT2 is found in mitochondria-associated membranes and has a mitochondrial targeting signal that promotes its association with mitochondria. *J Biol Chem.* 2009;284:5352–5361.
56. Kuerschner L, Moessinger C, Thiele C. Imaging of lipid biosynthesis: how a neutral lipid enters lipid droplets. *Traffic.* 2008;9:338–352.
57. Tan JS, Seow CJ, Goh VJ, Silver DL. Recent advances in understanding proteins involved in lipid droplet formation, growth and fusion. *J Genet Genomics.* 2014;41:251–259.
58. Minehira K, Gual P. Role of lipid droplet proteins in the development of NAFLD and hepatic insulin resistance. In: Valenzuela R, ed. *Non-Alcoholic Fatty Liver Disease—Molecular Bases, Prevention and Treatment.* IntechOpen; 2017.
59. Wang H, Bell M, Sreenivasan U, et al. Unique regulation of adipose triglyceride lipase (ATGL) by perilipin 5, a lipid droplet-associated protein. *J Biol Chem.* 2011;286:15707–15715.
60. Merchant SS, Kropat J, Liu B, Shaw J, Warakanont J. TAG, you're it! *Chlamydomonas* as a reference organism for understanding algal triacylglycerol accumulation. *Curr Opin Biotechnol.* 2012;23:352–363.
61. Li-Beisson Y, Beisson F, Riekhof W. Metabolism of acyl-lipids in *Chlamydomonas reinhardtii*. *Plant J.* 2015;82:504–522.
62. Park JJ, Wang H, Gargouri M, et al. The response of *Chlamydomonas reinhardtii* to nitrogen deprivation: a systems biology analysis. *Plant J.* 2015;81:611–624.
63. Liu B, Benning C. Lipid metabolism in microalgae distinguishes itself. *Curr Opin Biotechnol.* 2013;24:300–309.
64. Goodson C, Roth R, Wang ZT, Goodenough U. Structural correlates of cytoplasmic and chloroplast lipid body synthesis in *Chlamydomonas reinhardtii* and stimulation of lipid body production with acetate boost. *Eukaryot Cell.* 2011;10:1592–1606.
65. Welte MA. Expanding roles for lipid droplets. *Curr Biol.* 2015;25:R470–R481.
66. Walther TC, Farese Jr RV. Lipid droplets and cellular lipid metabolism. *Annu Rev Biochem.* 2012;81:687–714.
67. Szul MJ, Dearth SP, Campagna SR, Zinser ER. Carbon fate and flux in *Prochlorococcus* under nitrogen limitation. *mSystems.* 2019;4:e00254-18.
68. Kim J, Brown CM, Kim MK. Effect of cell cycle arrest on intermediate metabolism in the marine diatom *Phaeodactylum tricornutum*. *Proc Natl Acad Sci USA.* 2017;114:E8007–E8016.

69. Janssen JH, Wijffels RH, Barbosa MJ. Lipid production in *Nannochloropsis gaditana* during nitrogen starvation. *Biology*. 2019;8:5.
70. Wang X, Fosse HK, Li K, Chauton MS, Vadstein O, Reitan KI. Influence of nitrogen limitation on lipid accumulation and EPA and DHA content in four marine microalgae for possible use in aquafeed. *Front Mar Sci*. 2019;6:95.
71. Praveenkumar R, Shameera K, Mahalakshmi G, Akbarsha MA, Thajuddin N. Influence of nutrient deprivations on lipid accumulation in a dominant indigenous microalga *Chlorella* sp. *Biomass Bioenergy*. 2012;37:60–66.
72. Wang H, Zhang Y, Zhou W, Noppol L, Liu T. Mechanism and enhancement of lipid accumulation in filamentous oleaginous microalgae *Tribonema minus* under heterotrophic condition. *Biotechnol Biofuels*. 2018;11:328.
73. Zhu LD, Li ZH, Hiltunen E. Strategies for lipid production improvement in microalgae as a biodiesel feedstock. *Biomed Res Int*. 2016;2016:8792548.
74. Zhu JK. Abiotic stress signaling and responses in plants. *Cell*. 2016;167:313–324.
75. Li M, Yang L, Bai Y, Liu H. Analytical methods in lipidomics and their applications. *Anal Chem*. 2014;86:161–175.
76. Fuchs B, Süss R, Teuber K, Eibisch M, Schiller J. Lipid analysis by thin-layer chromatography—a review of the current state. *J Chromatogr A*. 2011;1218:2754–2774.
77. Fisk HL, West AL, Childs CE, Burdge GC, Calder PC. The use of gas chromatography to analyze compositional changes of fatty acids in rat liver tissue during pregnancy. *J Vis Exp*. 2014;85:51445.
78. Satomi Y, Hirayama M, Kobayashi H. One-step lipid extraction for plasma lipidomics analysis by liquid chromatography mass spectrometry. *J Chromatogr B*. 2017;1063:93–100.
79. Ishihara M, Kujiraoka T, Iwasaki T, et al. A sandwich enzyme-linked immunosorbent assay for human plasma apolipoprotein A-V concentration. *J Lipid Res*. 2005;46:2015–2022.
80. Li J, Vosegaard T, Guo Z. Applications of nuclear magnetic resonance in lipid analyses: an emerging powerful tool for lipidomics studies. *Prog Lipid Res*. 2017;68:37–56.
81. Li L, Han J, Wang Z, et al. Mass spectrometry methodology in lipid analysis. *Int J Mol Sci*. 2014;15:10492–10507.
82. Seppänen-Laakso T, Oresic M. How to study lipidomes. *J Mol Endocrinol*. 2009;42:185–190.
83. Furse S, Egmond MR, Killian JA. Isolation of lipids from biological samples. *Mol Membr Biol*. 2015;32:55–64.
84. Fujita A, Cheng J, Fujimoto T. Quantitative electron microscopy for the nanoscale analysis of membrane lipid distribution. *Nat Protoc*. 2010;5:661–669.
85. Abramczyk H, Surmacki J, Kopeć M, Olejnik AK, Lubecka-Pietruszewska K, Fabianowska-Majewska K. The role of lipid droplets and adipocytes in cancer. Raman imaging of cell cultures: MCF10A, MCF7, and MDA-MB-231 compared to adipocytes in cancerous human breast tissue. *Analyst*. 2015;140:2224–2235.
86. Jaeger D, Pilger C, Hachmeister H, et al. Label-free in vivo analysis of intracellular lipid droplets in the oleaginous microalga *Monoraphidium neglectum* by coherent Raman scattering microscopy. *Sci Rep*. 2016;6:35340.
87. Horn PJ, Ledbetter NR, James CN, et al. Visualization of lipid droplet composition by direct organelle mass spectrometry. *J Biol Chem*. 2011;286:3298–3306.
88. Zhu H, Fan J, Du J, Peng X. Fluorescent probes for sensing and imaging within specific cellular organelles. *Acc Chem Res*. 2016;49:2115–2126.
89. Lavis LD. Teaching old dyes new tricks: biological probes built from fluoresceins and rhodamines. *Annu Rev Biochem*. 2017;86:825–843.
90. Klymchenko AS. Solvatochromic and fluorogenic dyes as environment-sensitive probes: design and biological applications. *Acc Chem Res*. 2017;50:366–375.

91. Tatenaka Y, Kato H, Ishiyama M, et al. Monitoring lipid droplet dynamics in living cells by using fluorescent probes. *Biochemist.* 2019;58:499–503.
92. Maekawa M, Fairn GD. Molecular probes to visualize the location, organization and dynamics of lipids. *J Cell Sci.* 2014;127:4801–4812.
93. Daemen S, van Zandvoort M, Parekh SH, Hesselink M. Microscopy tools for the investigation of intracellular lipid storage and dynamics. *Mol Metab.* 2015;5:153–163.
94. Subramaniam HN, Chaubal KA. Evaluation of intracellular lipids by standardized staining with a Sudan black B fraction. *J Biochem Biophys.* 1990;21:9–16.
95. Aoki T, Hagiwara H, Fujimoto T. Peculiar distribution of fodrin in fat-storing cells. *Exp Cell Res.* 1997;234:313–320.
96. Koopman R, Schaart G, Hesselink MK. Optimisation of oil red O staining permits combination with immunofluorescence and automated quantification of lipids. *Histochem Cell Biol.* 2001;116:63–68.
97. Ohsaki Y, Shinohara Y, Suzuki M, Fujimoto T. A pitfall in using BODIPY dyes to label lipid droplets for fluorescence microscopy. *Histochem Cell Biol.* 2010;133:477–480.
98. Elle IC, Olsen LC, Pultz D, Rodkaer SV, Faergeman NJ. Something worth dyeing for: molecular tools for the dissection of lipid metabolism in *Caenorhabditis elegans*. *FEBS Lett.* 2010;584:2183–2193.
99. Greenspan P, Mayer EP, Fowler SD. Nile red: a selective fluorescent stain for intracellular lipid droplets. *J Cell Biol.* 1985;100:965–973.
100. BODIPY n.d. Thermo Fisher Scientific BODIPY 505/515. (accessed on 08 July 2019), Available from. https://www.thermofisher.com/order/catalog/product/D3921.
101. BODIPY n.d. Thermo Fisher Scientific BODIPY 493/503. (accessed on 08 July 2019), Available from https://www.thermofisher.com/order/catalog/product/D3922.
102. Elsey D, Jameson D, Raleigh B, Cooney MJ. Fluorescent measurement of microalgal neutral lipids. *J Microbiol Methods.* 2007;68:639–642.
103. Dutta AK, Kamada K, Ohta K. Spectroscopic studies of Nile red in organic solvents and polymers. *J Photochem Photobiol A.* 1996;93:57–64.
104. Brown MB, Miller JN, Seare NJ. An investigation of the use of Nile red as a long-wavelength fluorescent probe for the study of alpha 1-acid glycoprotein-drug interactions. *J Pharm Biomed.* 1995;13:1011–1017.
105. Collot M, Fam TK, Ashokkumar P, et al. Ultrabright and fluorogenic probes for multicolor imaging and tracking of lipid droplets in cells and tissues. *J Am Chem Soc.* 2018;140:5401–5411.
106. Spangenburg EE, Pratt SJ, Wohlers LM, Lovering RM. (2011). Use of BODIPY (493/503) to visualize intramuscular lipid droplets in skeletal muscle. J Biomed Biotechnol. 2011; article ID 598358:8:8.
107. Qian J, Tang BZ. AIE luminogens for bioimaging and theranostics: from organelles to animals. *Chem.* 2017;3:56–91.
108. Mei J, Hong Y, Lam JW, Qin A, Tang Y, Tang BZ. Aggregation-induced emission: the whole is more brilliant than the parts. *Adv Mater.* 2014;26:5429–5479.
109. Wang Z, Gui C, Zhao E, et al. Specific fluorescence probes for lipid droplets based on simple AIEgens. *ACS Appl Mater Interfaces.* 2016;8:10193–10200.
110. Wang E, Zhao E, Hong Y, Lam JWY, Tang BZ. A highly selective AIE fluorogen for lipid droplet imaging in live cells and green algae. *J Mater Chem B.* 2014;2:2013–2019.
111. Kang M, Gu X, Kwok RT, et al. A near-infrared AIEgen for specific imaging of lipid droplets. *Chem Commun.* 2016;52:5957–5960.
112. Jiang M, Gu X, Lam JWY, et al. Two-photon AIE bio-probe with large Stokes shift for specific imaging of lipid droplets. *Chem Sci.* 2017;8:5440–5446.
113. Niu G, Zhang R, Kwong JPC, et al. Specific two-photon imaging of live cellular and deep-tissue lipid droplets by lipophilic AIEgens at ultralow concentration. *Chem Mater.* 2018;30:4778–4787.

114. Zhuang W, Yang L, Ma B, et al. Multifunctional two-photon AIE luminogens for highly mitochondria-specific bioimaging and efficient photodynamic therapy. *ACS Appl Mater Interfaces*. 2019;11:20715–20724.
115. Qin W, Alifu N, Lam JWY, et al. Facile synthesis of efficient luminogens with AIE features for three-photon fluorescence imaging of the brain through the intact skull. *Adv Mater*. 2020;32:e2000364.
116. Park H, Li S, Niu G, et al. Diagnosis of fatty liver disease by a multiphoton-active and lipid-droplet-specific AIEgen with nonaromatic rotors. *Mater Chem Front*. 2021; 5:1853–1862.
117. Hamon N, Roux A, Beyler M, et al. Pyclen-based Ln(III) complexes as highly luminescent bioprobes for *in vitro* and *in vivo* one- and two-photon bioimaging applications. *J Am Chem Soc*. 2020;142:10184–10197.
118. Lepock JR. Cellular effects of hyperthermia: relevance to the minimum dose for thermal damage. *Int J Hyperth*. 2003;19:252–266.
119. Dugail I. Lysosome/lipid droplet interplay in metabolic diseases. *Biochimie*. 2014;96:102–105.
120. Dong H, Czaja MJ. Regulation of lipid droplets by autophagy. *Trends Endocrinol Metab*. 2011;22:234–240.
121. Hu R, Chen B, Wang Z, et al. Intriguing "chameleon" fluorescent bioprobes for the visualization of lipid droplet-lysosome interplay. *Biomaterials*. 2019;203:43–51.
122. Doskey CM, Buranasudja V, Wagner BA, et al. Tumor cells have decreased ability to metabolize H_2O_2: implications for pharmacological ascorbate in cancer therapy. *Redox Biol*. 2016;10:274–284.
123. Jiang G, Li C, Liu X, et al. Lipid droplet-targetable fluorescence guided photodynamic therapy of cancer cells with an activatable AIE-active fluorescent probe for hydrogen peroxide. *Adv Opt Mater*. 2020;8:2001119.
124. Shao A, Xie Y, Zhu S, et al. Far-red and near-IR AIE-active fluorescent organic nanoprobes with enhanced tumor-targeting efficacy: shape-specific effects. *Angew Chem Int Ed Engl*. 2015;54:7275–7280.
125. Lu H, Zheng Y, Zhao X, et al. Highly efficient far red/near-infrared solid fluorophores: aggregation-induced emission, intramolecular charge transfer, twisted molecular conformation, and bioimaging applications. *Angew Chem Int Ed Engl*. 2016;55:155–159.
126. Zhao Z, Su H, Zhang P, et al. Polyyne bridged AIE luminogens with red emission: design, synthesis, properties and applications. *J Mater Chem B*. 2017;5:1650–1657.
127. Wang D, Su H, Kwok RT, et al. Facile synthesis of Red/NIR AIE luminogens with simple structures, bright emissions, and high photostabilities, and their applications for specific imaging of lipid droplets and image-guided photodynamic therapy. *Adv Funct Mater*. 2017;27:1704039.
128. Zhang F, Li Z, Liu Y, et al. Rational construction of AIEgens with wide color tunability and their specific lipid droplet imaging applications. *J Mater Chem B*. 2020;8: 9533–9543.
129. Gu Y, Zhao Z, Su H, et al. Exploration of biocompatible AIEgens from natural resources. *Chem Sci*. 2018;9:6497–6502.
130. Reza AHMM, Zhou Y, Tavakoli J, Tang Y, Qin J. Understanding the lipid production mechanism in *Euglena gracilis* with a fast-response AIEgen bioprobe, DPAS. *Mater Chem Front*. 2021;5:268–283.
131. Ni JS, Liu H, Liu J, et al. The unusual aggregation-induced emission of coplanar organoboron isomers and their lipid droplet-specific applications. *Mater Chem Front*. 2018;2:1498–1507.
132. Ni JS, Lee MMS, Zhang P, et al. Swiss knife-inspired multifunctional fluorescence probes for cellular organelle targeting based on simple AIEgens. *Anal Chem*. 2019; 91:2169–2176.

CHAPTER SIX

AIE materials for lysosome imaging

Ankush Gupta[a,]*, Manoj Kumar[b], and Vandana Bhalla[b,]*
[a]Department of Chemistry, DAV University, Jalandhar, Punjab, India
[b]Department of Chemistry, UGC Sponsored Centre for Advanced Studies-II, Guru Nanak Dev University, Amritsar, Punjab, India
*Corresponding authors: e-mail address: chemankush@yahoo.co.in; vanmanan@yahoo.co.in

Contents

1. Introduction 145
   1.1 Design strategy and importance of AIE 146
   1.2 Design strategies of AIE molecular for lysosome 148
2. Lysosome viscosity tracing probes 148
   2.1 Lysosome tracking probes 156
3. Lysosome imaging for detection of other analytes 163
4. Conclusions 173
Acknowledgment 175
References 175

Abstract

The aggregation-induced emission (AIE) active bioprobes are known for their high photostability and extraordinary signal to noise ratio. In view of this, research efforts to synthesize new AIE bioimaging probes are at an incredible speed. In this chapter, we have summarized the various lysosome specific AIE active "turn-on" bioprobes having applications in lysosome imaging, monitoring of lysosome bioactivity and evaluation of their therapeutic effects. By discussing their design and operational mechanisms, we hope to provide more insight into designing new AIE bioprobes for specific sensing and imaging of lysosome having flexibility for broad range of biomedical applications.

1. Introduction

Lysosome, a vital organelles of eukaryotic cells,[1] contributes significantly toward maintaining various intracellular events such as autophagy, digestion and repair of plasma membrane, etc.[2,3] Lysosomal cell membrane has well-arranged enzymes, which have potential to break down all biopolymers such as proteins, nucleic acids, etc. The basic function of lysosome (pH 4.5–5.5) is to consume and eradicate nonfunctional bio-macromolecules.[4] To work properly for these functions, lysosome continuously changes its morphology and spatial distribution.[5] Abnormal biological activities of lysosome may cause severe health problems.[6–9] The proper functioning of lysosome is very important for the health of all the living systems. To attain information about cell bioactivities at cellular/molecular level for diagnosis and medical purposes, visualization and monitoring of lysosomal microenvironment is needed.

In view of this, research in this area has grown tremendously to explore the application of advanced diagnostic techniques including X-rays imaging, magnetic resonance imaging, ultrasound, acoustic radiation force impulse imaging, tomography, etc.[10] for visualization of lysosomal bioactivities. These techniques differ from each other in the way they generate images and deliver vital information. Although the use of all of these techniques is helpful but a few limitations such as less accuracy, poor image quality, prolonged volume of handling data, field distortion by radiations, hardware modifications, etc. limit their real time applications. Alternatively, fluorescence imaging is found to be among the indispensable technologies for molecular imaging, and use in preclinical research.[11,12] In comparison to other diagnostic techniques, fluorescence based imaging is advantageous due to its easy operation, high sensitivity, real-time imaging, no need of extra processing of data, short data acquisition time and high spatial resolution, etc.[13] Additionally, with fluorescence microscopy, image can be seen at micrometer or even nanometer levels.[14] In view of wonderful advantages of fluorescence imaging, researchers working in this area have developed a variety of fluorescence based probes using small organic molecules, conjugated polymers, nanoparticles, metal clusters, etc., for biological imaging applications.[15–20]

1.1 Design strategy and importance of AIE

The simple attachment of suitable binding groups to fluorophore can create organelle targeted systems. However, aggregation caused quenching (ACQ) inhibit the performance of most of the designed probes based on organic molecules.[21] To avoid the self quenching, low concentration of fluorescent probe is applied, which adversely affects signal-to-noise ratio, photo bleaching and phototherapy outcome.[22] Tang et al. reported several AIE active silole derivatives which are non fluorescent in their dispersed form but showed strong fluorescence upon aggregation. The system having AIE feature shows exceptional a distinctive property that is contrary to ACQ. In aggregated state, the inhibition of intramolecular rotations due to physical stacking leads to bright and stable emission (Scheme 1).[23] Tang et al. proposed restriction in intramolecular rotation (RIR) is the key operating pathway for AIE phenomenon.[24] Besides RIR, various other mechanisms such as J aggregation, excited state intramolecular proton transfer (ESIPT), twisted intramolecular charge transfer (TICT), etc., are also prevalent.[25,26]

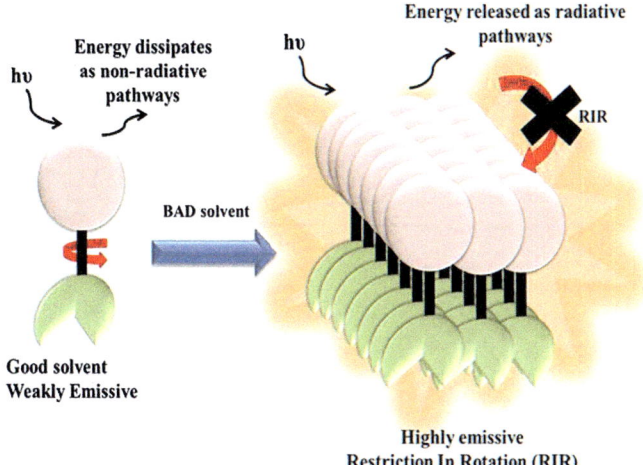

Scheme 1 Scheme showing AIE phenomenon.

AIE materials are important building blocks for the preparation of "lighted" sensors for various molecular recognition and imaging applications. The emissive aggregates of AIE active materials provide additional advantages of selectivity, sensitivity, and operational simplicity.[27] Further, the AIE-active probes are environmentally stable, synthetically readily accessible and exhibit "turn-on" response toward target analyte. Due to their highly emissive nature, AIE active probes are suitable to use at higher concentrations with significant signal reliability and are resistant to photobleaching in comparison to that of ACQ derivatives.[28] As a result, AIE active materials are extensively employed for long-term cell tracking in various cell organelle,[29–32] photodynamic therapy and chemotherapy.[33–35]

In this chapter, we have summarized various AIE active probes, which are designed to mark the lysosomes for imaging and therapeutic purposes.

1.2 Design strategies of AIE molecular for lysosome

The design of a lysosome-specific fluorescent bioprobe is based on two strategies. The first strategy aims at designing bioprobe having affinity for lysosomal enzymes. The probe is non fluorescent and becomes highly emissive upon reaction with cellular lysosomal enzymes (Scheme 2). In the second strategy, the binding sites of the probe are protected by weak bases (*N*, *N*-dimethyl amine or morpholine), which upon penetration into the cell undergo deprotection in acidic lysosomal environment to generate emissive aggregates (Scheme 2). The specific targeting and biological reaction in lysosomes completely avoids the interference from the other non-target analytes, which maximizes the specificity of light-up probe.[36–38]

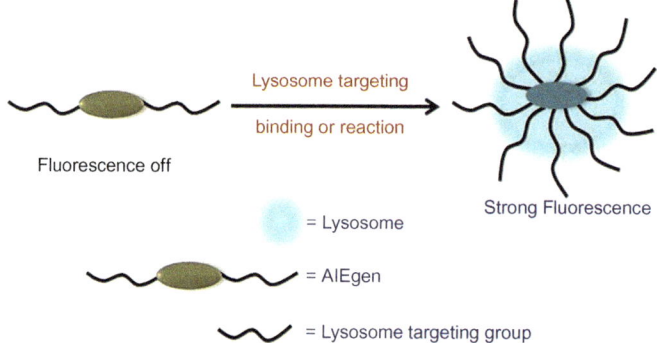

Scheme 2 Scheme showing lysosome target "Turn-on" AIE probes.

Based on these design strategies, several AIE derivatives have been designed for theranostic and bioimaging applications. Thus, lysosomal light-up AIEgens and their applications have been outlined. Through discussion of the summarized operational mechanisms of these probes, we offer a platform for development of advanced AIEgens having high selectivity, sensitivity and flexibility for broader range of biomedical applications.

2. Lysosome viscosity tracing probes

The pH and viscous environment of lysosome[39,40] significantly affect[39–41] various biological activities such as signal transduction, transportation and metabolite diffusion, etc.[42–44]

In view of this, Tang et al. developed lysosome targetable piperazine appended AIE active tetraphenylethane (TPE) derivative **1**, which showed pH and viscosity dependent changes in its emission intensity behavior. The piperazine unit was attached to the TPE scaffold to provide water compatibility to the fluorescent TPE unit.[2] In comparison to commercial available LysoTracker Red (LTR) dye, the designed derivative **1** showed negligible cytotoxicity, good tracing ability and better photostability. The derivative **1** exhibited insignificant change in fluorescent quantum yields in the pH range 1 to 7. In alkaline environment (pH between 7 and 13), quantum yield is gradually increased (12.7%) (Fig. 1). The lowest quantum yield observed in acidic media in case of derivative **1** is attributed to the enhanced water solubility of its protonated form, which increased the intermolecular repulsion. In basic solution, increase in quantum yield of derivative **1** is due to formation of fluorescent aggregates due to its AIE characteristics.

1

The quantum yield of derivative **1** was found to be highly sensitive to the viscosity. For cell imaging, live Hela cells were stained with TPE derivative **1** and Lyso Tracker Red (Fig. 2). The correlation coefficient of 0.82 confirmed the specifically localization of **1** in the cell lysosomes.

Localization of derivative **1** in lysosomes is due to the co-presence of piperazine groups in molecule. The probe was further utilized to demonstrate lysosomal migration using chloroquine, a typical lysosomal drug in

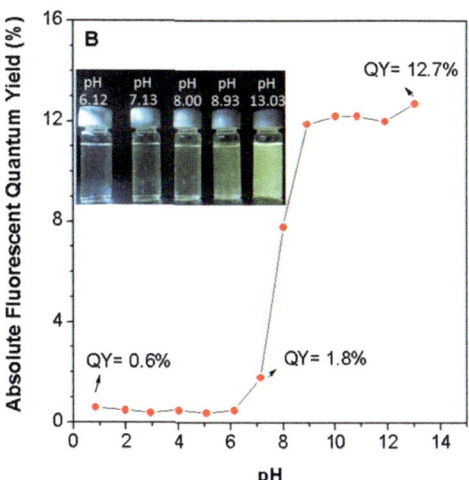

Fig. 1 Quantum yield of **1** in buffer solutions ($\lambda_{ex}=360$ nm). Insert are images of **1** in buffer solutions. *Copyright 2017, Royal Society of Chemistry.*

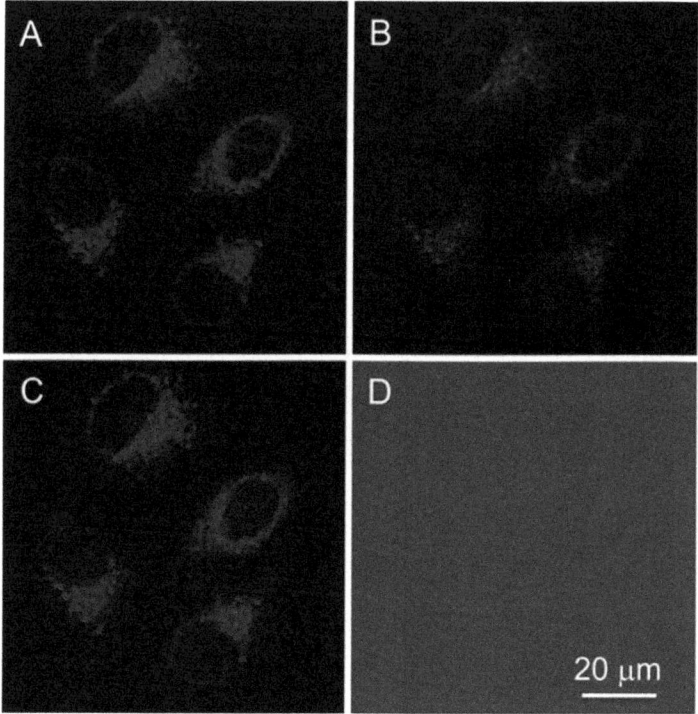

Fig. 2 (A) and (B) are the confocal images of HeLa cells stained with **1** and LTR. (C) Merged image. (D) Bright field images. *Copyright 2017, Royal Society of Chemistry.*

the HeLa cells. The confocal images of cells incubated with **1** upon stimulating with chloroquine show lysosomal movement without disturbing to the cells (Fig. 3).

A lysosomal viscosity mapping AIE active probe **2**, based on p-hydroxybenzylideneimidazolidin one has been reported.[45]

2

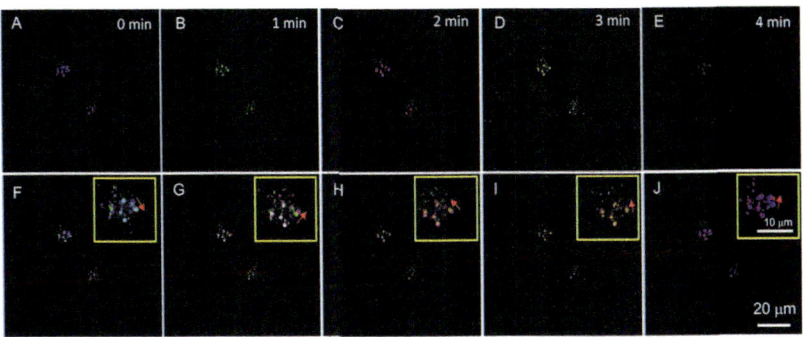

Fig. 3 Confocal images of HeLa cells with **1** and chloroquine for (A–E) 0 min, 1 min, 2 min, 3 min and 4 min, respectively. Merged images (F–J) 0 and 1 min, 1 and 2 min, 2 and 3 min, 3 and 4 min, 0 and 4 min, respectively. Inset images: arrow indicates direction of movement of lysosome. *Copyright 2017, Royal Society of Chemistry.*

AIEgen **2** showed high emission at longer wavelength in comparison to *p*-hydroxybenzylideneimidazolidin
one analogue due to increase in conjugation. Further, phenol groups were modified to introduce two morpholine groups, which worked as antenna for Colocalization of lysosomes in living cells. The presence of twisted linkages between imidazolidinone and phenolic units quenched the fluorescence of derivative **2**. However, with rise in viscosity of the solution, the twisting of the bonds is inhibited. Further, the fluorescence response of AIEgen **2** with increased viscosity in different pH values is studied. It is found that, at given viscosity and change in pH did not bring any significant variation in the fluorescence of derivative **2** (Fig. 4).

Based on these results, live cell imaging of MCF-7 cells was performed to track lysosome using confocal microscope (Fig. 5). The significant overlap coefficient of about 0.934 was observed between AIEgen **2** and LTR labeled signal, which confirmed lysosomal localization of **2**.

The lysosomal viscosity mapping potential of AIEgen **2** was further demonstrated in MCF-7 cells. The experiment confirmed that change in emission intensity was mainly from variation in intracellular lysosomal viscosity (Fig. 6).

Jiang et al. also reported fluorescent probes **3** and **4**, for measuring viscosity changes in lysosomes.[46] The indole ring was incorporated in probe **3** to enhance its lysosome target ability. AIEgen **3** was found to be pH-independent (pH range 4.0–8.0) at different viscosity levels. However,

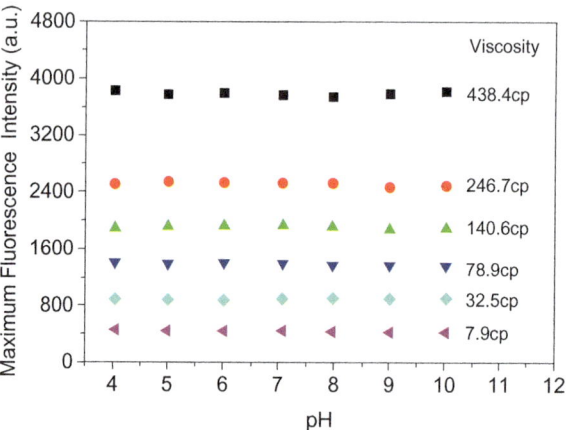

Fig. 4 Maximum fluorescence intensity of **2** with pH variation at different viscosities. *Copyright 2018, Royal Society of Chemistry.*

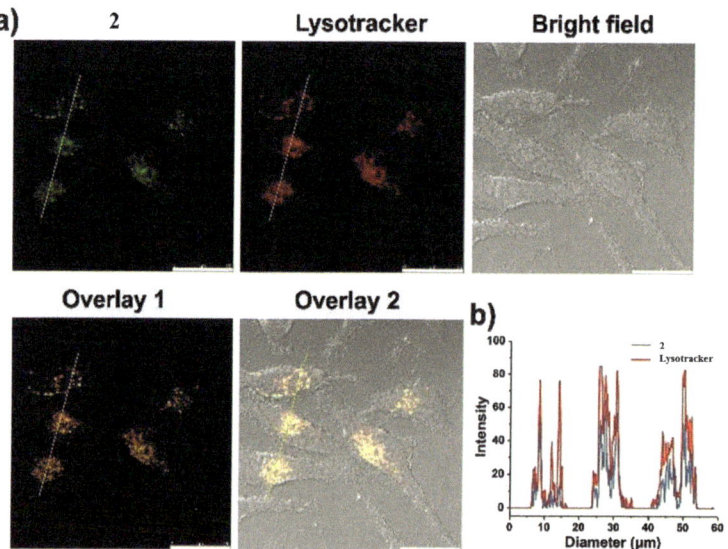

Fig. 5 Confocal images of treated MCF-7 cells. (A) Imaging of **2** with LTR Green image: **2** stained channel 1. Red image: Lyso-Tracker labeled channel 2 (B) Intensity profile through MCF-7 cells. *Copyright 2018, Royal Society of Chemistry.*

derivative **4** showed pH dependence emission changes. At lower pH the emission changes in case of AIEgen **4** are attributed to the protonation of **4**, which influenced the intramolecular charge transfer (ICT) in the system. On the contrary, in case of **3**, insignificant pH dependence was observed due to the presence of relatively acidic indole group. Thus, due to its better specificity, AIEgen **3** showed potential microviscosity probe.

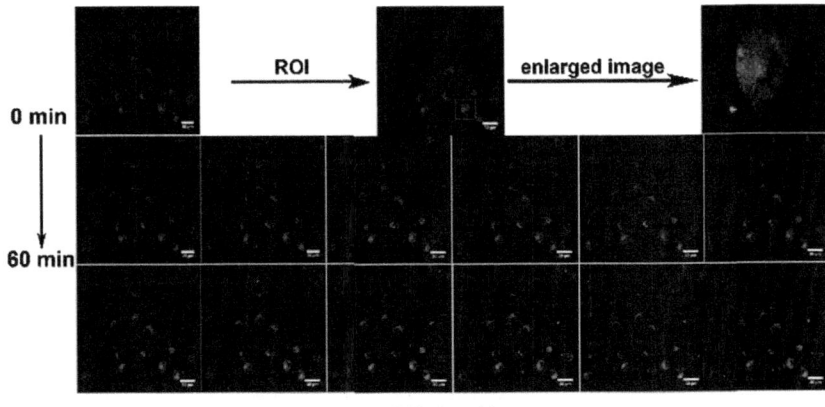

Fig. 6 Fluorescence changes in **2** stained MCF-7 cells and after exposure to dexamethasone. *Copyright 2018, Royal Society of Chemistry.*

The AIEgen **3** showed 20-fold enhanced emission in viscous solvent (glycerol fraction from 0% to 99%) (Fig. 7, A–B). Lysosomal viscosity was examined in HeLa cells, which were sequentially incubated with LTR. The overlap of probe and LTR emission signals was found to be 92% (Fig. 7, C1–C4), which suggests excellent lysosomal targeting affinity of probe **3**. The time dependent imaging of live cells after treatment with dexamethasone confirms the lysosomal movement (Fig. 7, D1–D4). Furthermore, the fluorescent viscosity probe **3** was also used to examine enhanced viscosity of lysosomes throughout the mitophagy process (Fig. 8).[6]

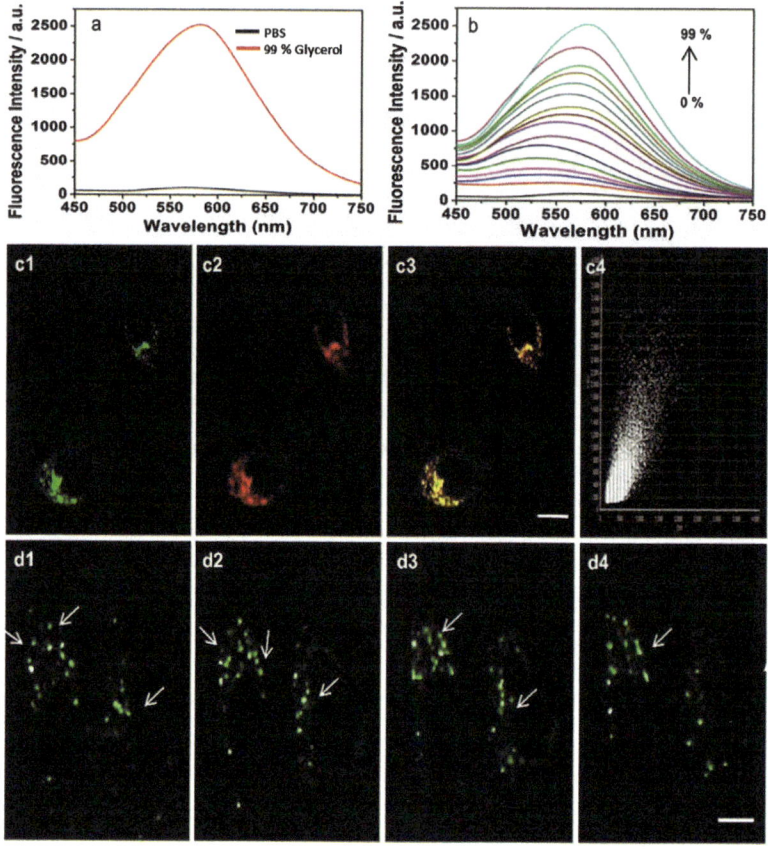

Fig. 7 (A) Fluorescence spectra of **3** in PBS:glycerol. (B) Viscosity dependent emission spectra (C1–C4) Colocalization of **3** with LTR. (D1–D4) Time dependent cell images after treatment with dexamethasone. *Copyright 2018, American Chemical Society.*

Tang et al. designed and synthesized TPE based AIEgen **5** having coumarin moiety,[47] to investigate the pH change in organelles. The fluorescence spectrum of **5** exhibited fluorescence at 435 nm in 99% aqueous solution, however, upon changing pH from 1 to 4, significant rise in florescence of **5** was noted. On the contrary, decrease in the fluorescent intensity was detected in pH range 4 to 13 (Fig. 9). At low pH (from 1 to 4) protonation of imine groups results in the formation of protonated species having good solubility. However, at pH 4, the degree of protonation of **5** is decreased and due to enhance hydrophobic clustering, an increase in fluorescence emission is detected. At higher pH (up to 13), the π-π stacking of molecules of coumarin moiety is enhanced and hence, emission intensity of the aggregated species is decreased (Fig. 9B).

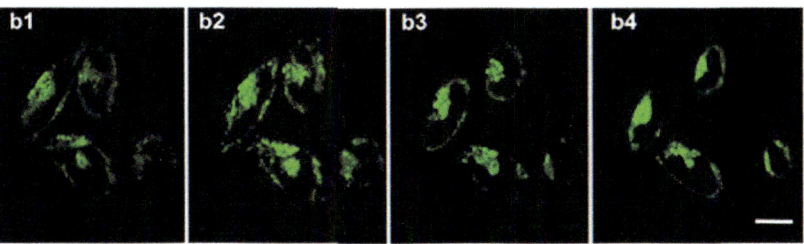

Fig. 8 (b1–b4) Time dependent confocal images of restained cells (0 min., 30 min., 60 min. and 120 min.), during starvation-induced mitophagy. *Copyright 2018, American Chemical Society.*

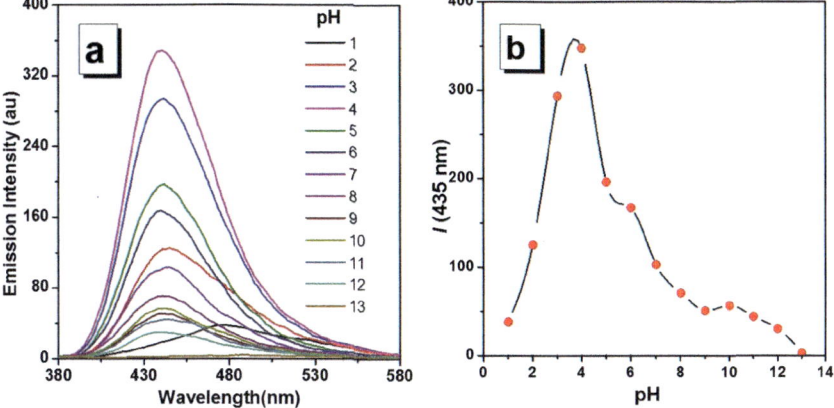

Fig. 9 (A) Fluorescence spectra of **5** at different pH. (B) Emission intensity (I) plot at different pH. *Copyright 2016, Royal Society of Chemistry.*

5

Co-localization imaging experiments in HLF, MCF-7 and Hela cell lines incubated with **5**, Lyso Tracker Green DND-26, and LysoTracker Red DND-99 showed different emission with variable emission intensity in all the channels (Fig. 10). Further studies also showed the excellent photostability and biocompatibility of AIEgen **5** in comparison to other commercially available dyes.

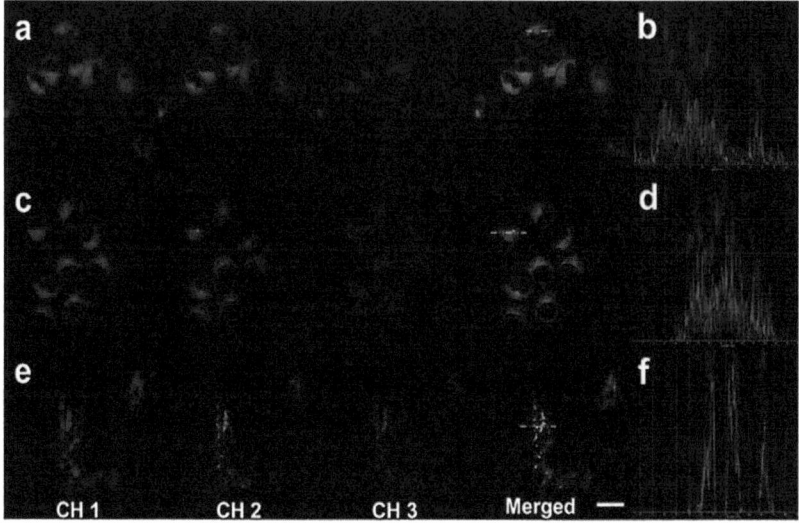

Fig. 10 Confocal images of **5**, LTG DND-26 and LTR DND-99 with (A) Hela, (B) MCF-7, (C) HLF cells (B,D and F) corresponding intensity profile of regions of interest. Blue, green and red lines are for **5**, LTG DND-26 and LTR DND-99, respectively. CH 1: 5, Blue Channel; CH 2: LTG DND-26, Green Channel; CH 3: LTR DND-99, Red Channel. All are merged images. Scale, 20 μm. *Copyright 2016, Royal Society of Chemistry.*

2.1 Lysosome tracking probes

The need for water compatibility of the probe to attain low background signal and "turn-on" optical response restricts the use of hydrophilic substrate as the scaffold for designing these probes.[23] Furthermore, modifications required in the probe for detection of hydrophobic elements are complexed and tedious. The discovery of AIE phenomenon has provided new platform to create "turn-on" probes for targeted enzyme recognition.[48–50] Tang et al. designed and synthesized the first AIE-ESIPT based lysosome targeting bioprobe **6** having morpholines as a lysosome targeting units and acetoxyl groups as esterase binding substrate sites.[31] Lysosome enzyme detection was facilitated by the presence of hydrophobic and hydrophilic units in derivative **6**. Initially, presence of protecting acetyl groups hindered the hydrogen bonds and free rotation of the N—N bond, hence, emission is quenched. Eventually, the generation of the hydroxyl groups by esterase hydrolysis facilitated the intramolecular hydrogen bonding, which results in "turn-on" response owing to both ESIPT and AIE phenomenon (Scheme 3).

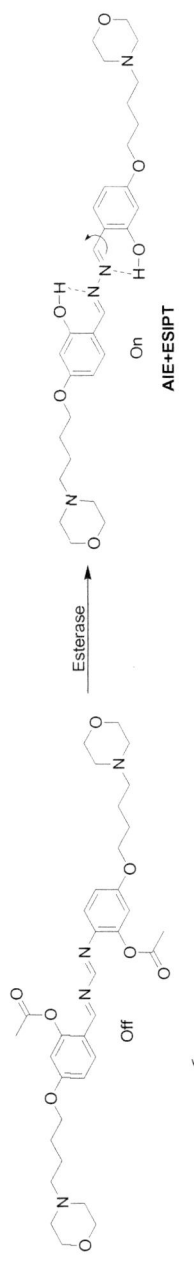

Scheme 3 Reaction of **6** with esterase.

In MCF-7 cells, upon incubation of the probe **6** it is taken by the lysosome under the direction of morpholine groups. After incubation, fluorescence emission signals were obtained from lysosomes, which overlapped well with signal of the commercially available LTR, which indicated the high specificity of probe **6** toward lysosome (Fig. 11A–D). Further, the increase in fluorescence intensity over the time, revealed the potential of probe to in situ track the esterase activity.

The potential of probe **6** was also revealed for tracking lysosomal movements continuously by confocal microscopy (Fig. 12). The merged pictures collected at different times clearly showed the slight movement of lysosomes by probe

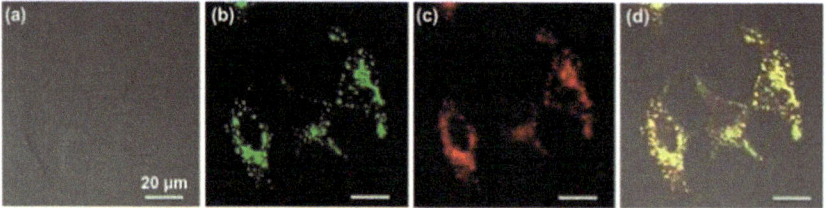

Fig. 11 (A) Confocal bright-field image of MCF-7 cells with **6** and LTR (B) channel 1 for **6**; (C) channel 2 for LTR; and (D) Merged image of (B) and (C). *Copyright 2014, Royal Society of Chemistry.*

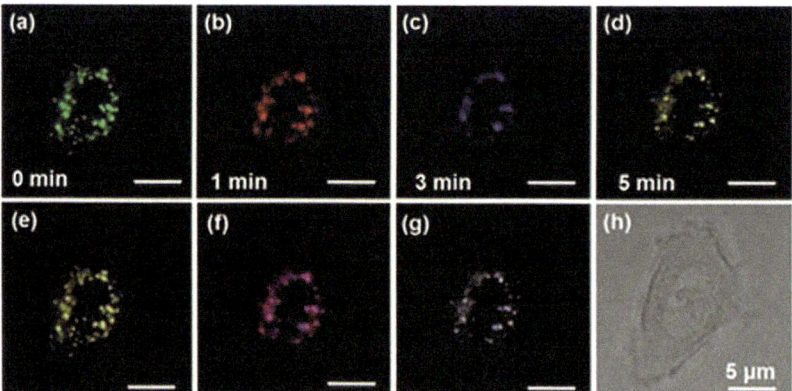

Fig. 12 Confocal images of MCF-7 cell with **6** stimulated with chloroquine (A–D) images at different stimulation times. (E–G) Merged images at: (E–H) 0 and 1 min., 1 and 3 min., 3 and 5 min., respectively, (H) bright-field image. Scale = 5 mm. *Copyright 2014, Royal Society of Chemistry.*

6 (Fig. 12E–G). Further, 3-(4,5-dimethylthiazol-2-yl)-2,5-diphenyltetrazolium bromide (MTT) assays confirmed the biocompatibility of probe **6**.

The AIE-active probe **6** also showed selective response toward esterase in the presence of different interfering species. The esterase showed 70–200 fold increase in emission intensity in comparison to other interfering substances. The detection limit of probe **6** for esterase was found to be of 2.4×10^{-3} UmL^{-1}.

Building on the same lines, to investigate of autophagy process, a new lysosome specific bioprobe **7** was developed having AIE-ESIPT characteristics. In the bioprobe **7**, the morpholine group served as the lysosome targeting ligand.[51] The good biocapability of the probe was revealed by MTT cell proliferation assay. Lysosome-targeting performance of bioprobe **7** was investigated by fluorescence microscopic technique.

The overlapped yellow emission of bioprobe **7** with the red fluorescence from LTR confirmed the selectivity of **7** for lysosomes in living cells (Fig. 13).

The continuous scanning of MCF-7 cells treated with bioprobe **7** showed negligible change in the fluorescence emission, whereas, emission intensity of LTR was decreased to 50%, which confirms better photostability of bioprobe **7** in comparison to commercially available LTR. The bioprobe **7** upon incubation with HeLa cells and upon treatment with rapamycin for different periods of time provided intracellular activity autophagy (Fig. 14A–E).

Fig. 13 (A) Confocal image of HeLa cells stained with **7**. (B) LTR (C) Merged image of (A) and (B). *Copyright 2015, Wiley-VCH.*

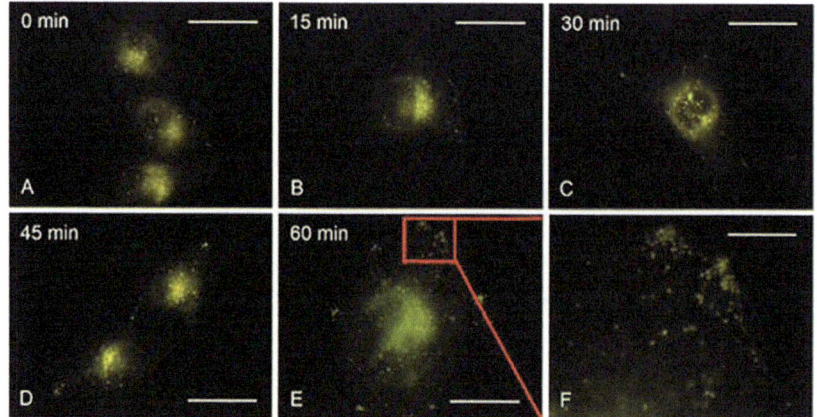

Fig. 14 (A–E) Confocal images of HeLa cells with **7** before and after rapamycin treatment at different periods of time. Scale for (A)–(E) 30 μm (F) Enlarged image of (E). *Copyright 2015, Wiley-VCH.*

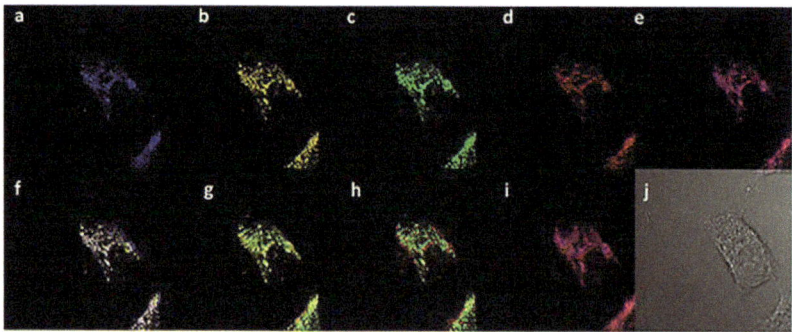

Fig. 15 Confocal images of an L929 cell stained with **8** upon stimulation with chloroquine. Images at time (A–E) 0 min, 5 min, 10 min, 15 min and 20 min, respectively. Merged images at: (F–I) 0 and 5 min, 5 and 10 min, 10 and 15 min, 15 and 20 min, respectively. (J) bright-field image. *Copyright 2016, Elsevier.*

Liu et al. examined the lysosome targeting ability of highly photostable probe **8**.[52] The confocal microscopic images of probe **8** in chloroquine treated L929 cell lines showed the movements of lysosomes in cells (Fig. 15). The chloroquine induced the lysosomal movement without disturbing the cells.[53]

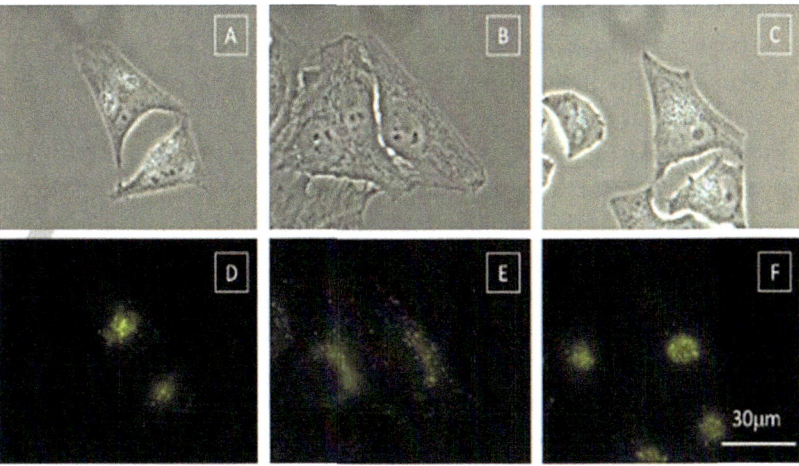

8

The negligible movement of lysosomes was clearly observed in the merged images with time interval (Fig. 15). In addition, AIEgen **8** was also successfully applied for diagnosis of apoptosis process.

AIE active lysosome specific bioprobe **9** was employed to monitor the dynamic lysosomal changes including tracking of endocytosis process of macromolecules.[5] In synthesized AIEgen **9**, the presence of morpholine units permitted the derivative **9** to easily target the lysosomes.

The autophagy process was monitored by fluorescence microscopic images. The lysosomes of HeLa cells were clearly visualized as green spots (Fig. 16) and portion of lysosome was increased after cells were treated with rapamycin (Fig. 16E). This study reveals the real time application of probe **9** to monitor regulating in autophagy process in HeLa cells. Further, upon

Fig. 16 Bright field images of HeLa cells stained with **9** (A–C). before treatment (A, D), after treatment of rapamycin (B, E) and after treatment of rapamycin and 3-MA (C, F). *Copyright 2019, Wiley-VCH.*

using 3-Methyladenine (3-MA), no change in amount and location of lysosomes was observed in 30 min. This result showed that despite the presence of rapamycin, the initial stage of autophagy process was inhibited.

9

To enhance the biocapability of probe, a near-infrared (NIR) AIEgen **10** was synthesized.[54]

10

The complex **10** showed formation of amorphous particles in high water content, indicating instability of the formed aggregates. However, in the presence of biocompatible Pluronic F127, the stability of nanoparticles was enhanced.[55] Interestingly, the as prepared **10@F127** was found to be stable under lab conditions for few weeks. The MTT assay experiment showed more than 96% cell viability of nanoaggregates of **10@F127**. Further, microscopic images of HeLa cells incubated with **10@F127** showed the internalization of aggregates by living cells (Fig. 17). The overlap coefficient of HeLa cells stained with **10@F127** and Lyso Tracker Green (LTG) was found to be about 88.3%, which indicated that the aggregates of **10@F127** could efficiently localize in the lysosomes.

The lysosome-specific localization property of **10@F127** is attributed to acid-base properties of terpyridyl complexes. Upon internalization of the ensemble by endocytosis, **10@F127** probe molecules moved to lysosomes where these complexes were protonated and trapped inside the lysosomes.

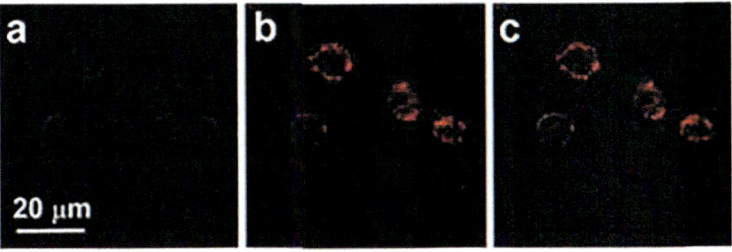

Fig. 17 Confocal images of HeLa cells stained with (A) LTG and (B) **10@F127**. (C) The merged image. *Copyright 2018, Royal Society of Chemistry.*

3. Lysosome imaging for detection of other analytes

Lysosomal endogenous hydrogen sulfide (H_2S) is found to induce cell death through destabilization of its cell wall and release of lysosomal proteases.[56,57] Jian et al. reported development of AIE dots for selective detection of lysosomal H_2S.[58] Simple precipitation method was used to prepare AIE dots by mixing amphiphilic block copolymer **11** and compound **12**. The generated AIE dots were stable for more than four months at room temperature. The dilute solution of AIE dots showed weak fluorescence owing to photo-induced electron transfer (PET). However, addition of H_2S showed significant increase in emission intensity due to cleavage of ester linkage in assemblies of derivative **12** to release AIE active compound **13**. Various experimental studies confirmed the cleavage of ester linkage.

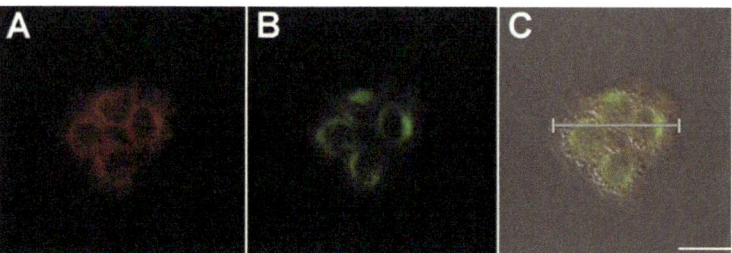

Fig. 18 (A) Confocal images of HeLa cells with AIE Dots and H$_2$S; (B) HeLa cells with LTG; (C) Merged image. *Copyright 2017, Royal Society of Chemistry.*

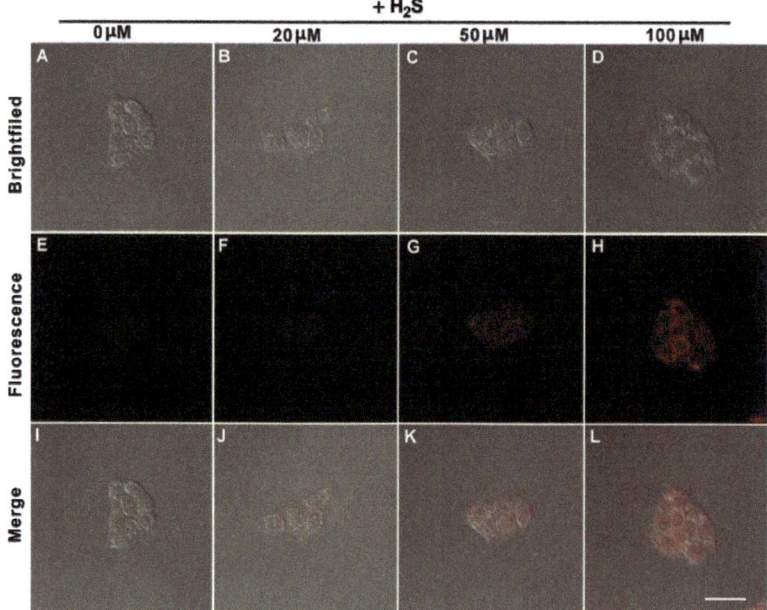

Fig. 19 Confocal images of HeLa cells stained with AIE dots and then different concentration of H$_2$S (A, E and I for 0 μM, B, F, J for 20 μM, C, G, K for 50 μM and D, H, L for 100 μM). *Copyright 2017, Royal Society of Chemistry.*

Highly co-localized fluorescence was observed when HeLa cells co-incubated with AIE dots, H$_2$S and LTG (Fig. 18). The highly fluorescent confocal images confirmed the internalization of AIE dots in lysosomes and their applications for imaging H$_2$S in living cells.

The as prepared AIE dots were further explored for investigating exogenous H$_2$S levels in HeLa cells. The AIE dots treated cells showed an enhancement in red fluorescence with increasing H$_2$S concentration from 20 to 100 μM (Fig. 19).

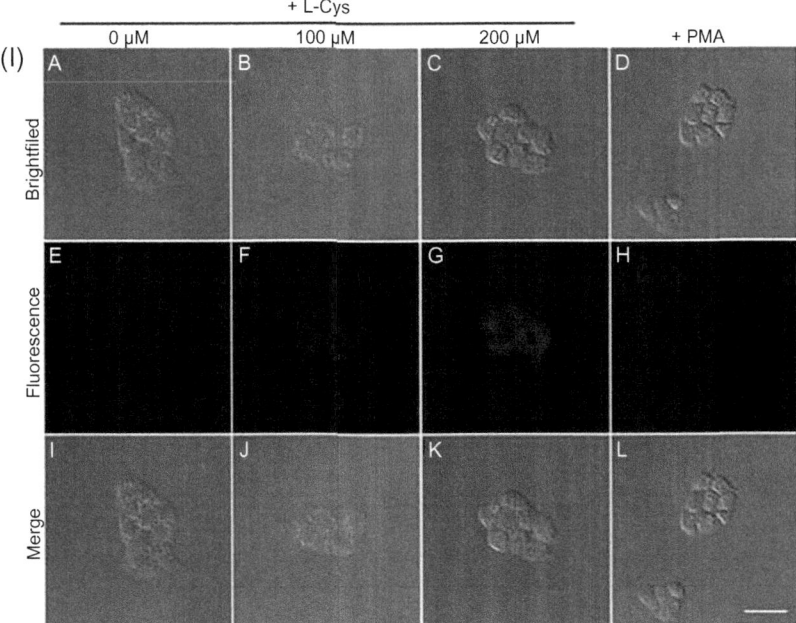

Fig. 20 Confocal images of HeLa cells stained with different concentration of L-Cys (A, E, I for 0 µM; B, F, J for 100 µM; C, G, K for 200 µM and D, H, L for PMA) and then incubated with AIE dots. *Copyright 2017, Royal Society of Chemistry.*

Further, in live cells endogenous H_2S was visualized by using AIE dots. It was found that free AIE dots treated cells showed no fluorescence emission, however, pre-treated cells with L-Cys showed strong red fluorescence (Fig. 20). This result showed the ability of AIE dots to detect endogenous H_2S in lysosomes in presence of external agent L-Cys.

Building on same strategy, Chen et al.[59] developed ratiometric lysosomal specific AIE dots for imaging of endogenous lysosomal hypochlorous acid (HClO) in live cells (Scheme 4).

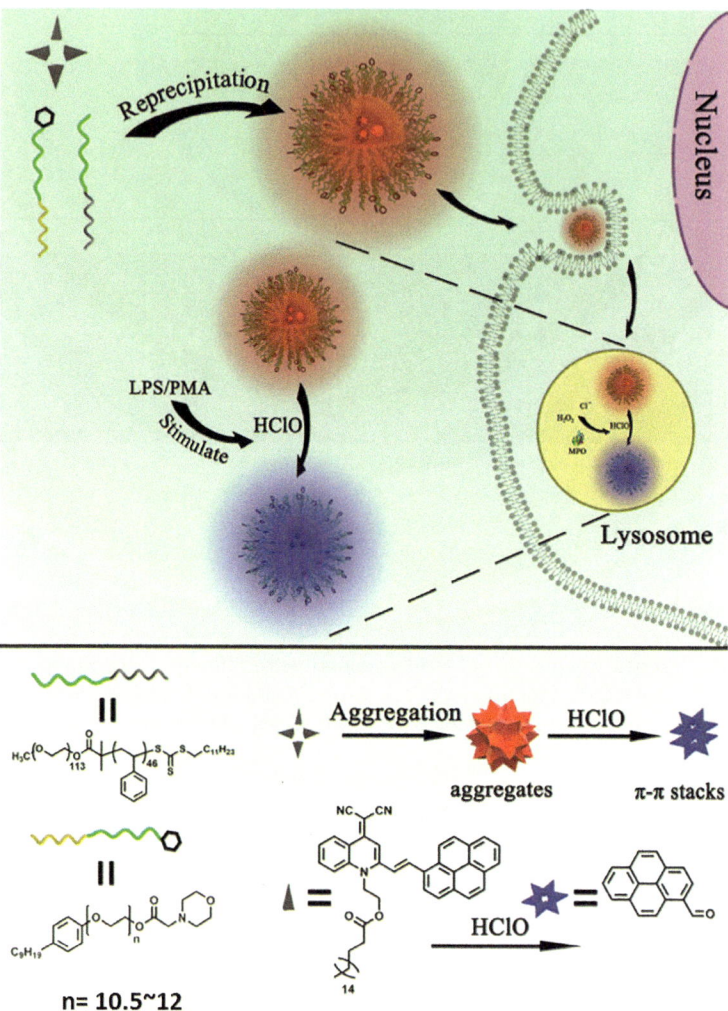

Scheme 4 AIE dots for the detection of HClO. Copyright 2018, Royal Society of Chemistry.

14

15, n= 10.5-12

16

The coprecipitation strategy was used to prepare AIE dots by employing lysosome-targetable surfactant **15**, an amphiphilic block copolymer **16** and HClO-responsive probe **14**. The as prepared AIE dots showed good biocompatibility to monitor the endogenous HClO in lysosomes. The pretreated HeLa cells (Fig. 21E, I and M), showed gradual increase in blue fluorescence (Fig. 21J, K and L) and a decrease in the red emission upon exposure to varied portions of HClO (Fig. 21F, G and H).

As a result, ratio of emission intensity (blue/red) was found to increase significantly (Fig. 21N, O and P). Similarly, the endogenous HClO in living cells was also imaged using AIE dots.[60] As shown in Fig. 22, after incubation with only AIE Dots, the RAW264.7 cells displayed red emission with low ratio of intensity in blue as well as red region (blue/red) (Fig. 22D, G and J). After stimulating the cells with lipopolysaccharide (LPS) and phorbol myristate acetate (PMA), a significant decrease in red fluorescence and increase in blue fluorescence was observed (Fig. 22E, H and K). Further, the cells incubated with LPS, PMA and N-acetylcysteine (NAC) showed restoration of red emission (Fig. 22F, I and L), which confirmed the role of AIE dots as a ratiometric nanoprobe.

More recently, AIEgens have been designed which exhibit high potential application in non-invasive treatment of malignant tissue/cells.[61] Recently, AIEgens **17** and **18** are found to exhibit photosensitizing characteristics for specific lysosomal photodynamic therapy.[62] Both the AIEgens

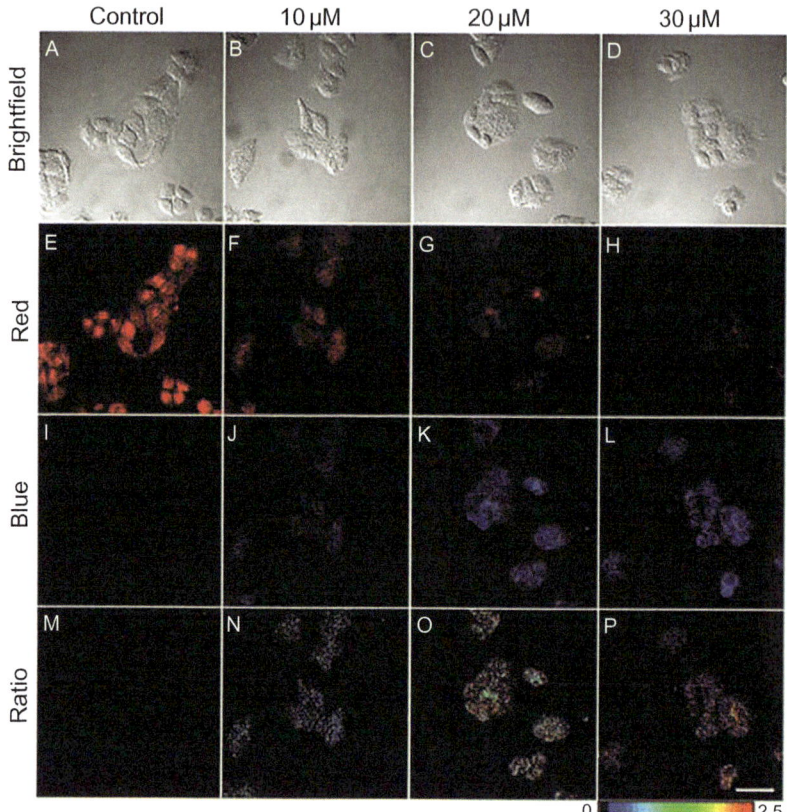

Fig. 21 Confocal images of HeLa cells stained with AIE dots with the addition of different concentration of HClO. The ratio image is created by dividing intensity of blue fluorescence to red fluorescence. *Copyright 2018, Royal Society of Chemistry.*

were found to be good biocompatible and stained lysosome in live cell with high specificity, even without washing step. Though both the AIEgens showed $^1O_2$ generation capability but better photosensitization was reported in case of AIEgen **18**.

The efficient properties of both the AIEgens to generate $^1O_2$ led to the ablation of lysosomal cancer cells through a targeted photodynamic therapy. The HeLa cells pretreated with AIEgen **17/18** and 2′,7′-dichlorofluorescein diacetate are exposed to white light (Fig. 23A).

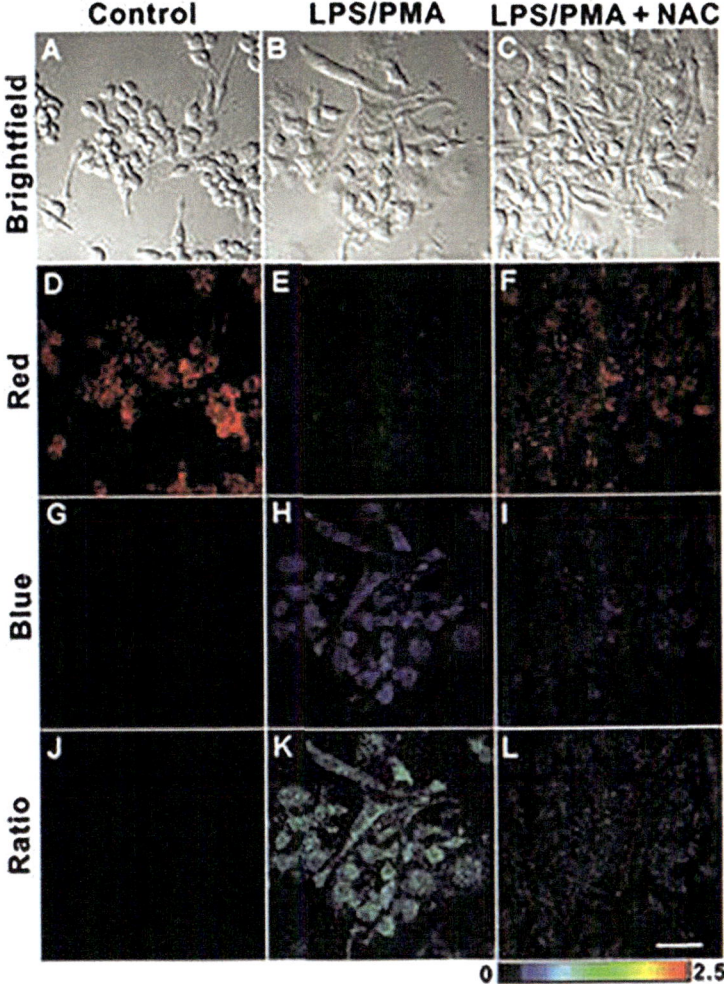

Fig. 22 (A, D, G) Confocal images of RAW264.7 cells incubated with AIE Dots. (B, E, H) incubated with LPS, after that co-incubation with PMA and AIE dots for 3 h. (C, F, I) incubated with LPS and NAC, after that co-incubation with PMA and AIE Dots (J, K, L). The ratio image is created by dividing intensity of blue fluorescence to red fluorescence. Copyright 2018, Royal Society of Chemistry.

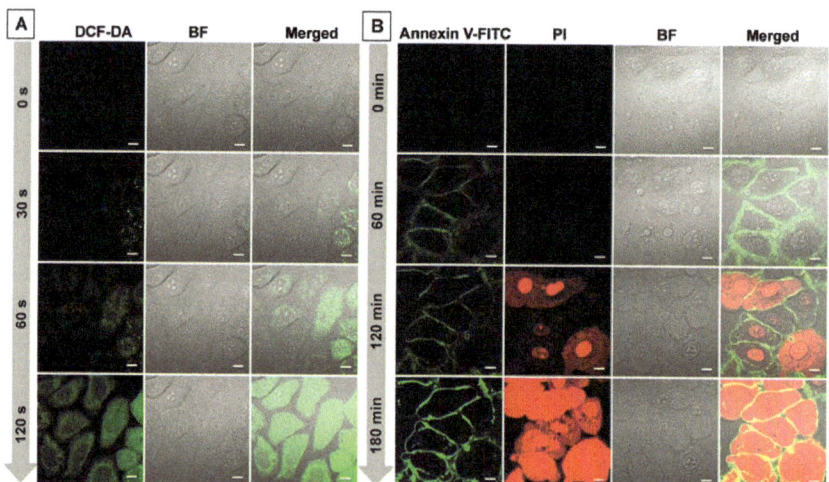

17, X= O
18, X= C

The pretreated cells showed the bright green fluorescence from cell cytoplasm owing to oxidation of 2′,7′-dichlorofluoresceindiacetate to 2′,7′-dichlorofluorescein via endogenously produced singlet oxygen. Further, the cell apoptosis was monitored and it was found that upon irradiating the cells with white light, the cell morphology is significantly disturbed and in addition to green fluorescence, red intense fluorescence is observed (Fig. 23B), which indicates cell death.

Fig. 23 (A) Confocal Images of Hela Cells stained with **18** under white light irradiation (B) Cell apoptosis observed after treated with **18** under white light irradiation. *Copyright 2020, Royal Society of Chemistry.*

Further, Wang et al.[63] reported a Lysosomal β-N-acetylhexosaminidase (Hex) specific fluorescent probe **19**. The N-acetyl-β-D-glucosaminide moiety (as a Hex-activatable group) is combined to TPE scaffold through pyridinium spacer. The probe **19** exhibits water solubility and weak emission in the biological environment. The generation of phenolate intermediate via Hex-catalyzed hydrolysis (catalytic release of N-acetylhexosamine) of probe **19** led to generation of derivative **20**. Derivative **20** due to its poor water solubility formed highly fluorescent aggregates (Scheme 5).

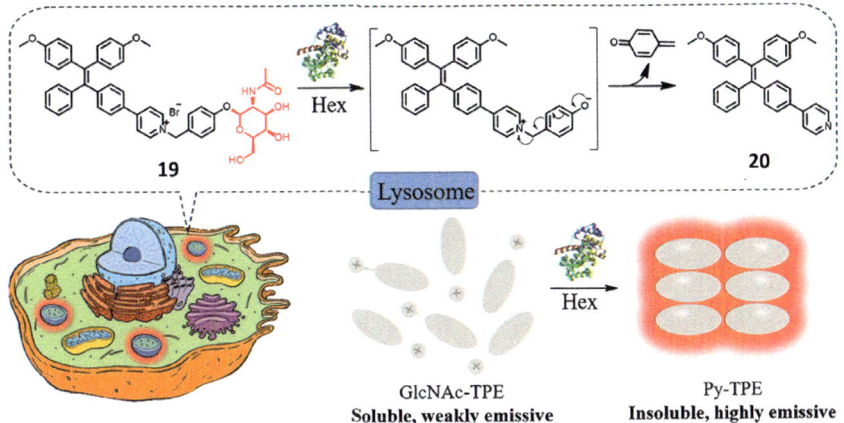

Scheme 5 Detection mechanism of **19** toward Hex. Copyright 2019, American Chemical Society.

The human colorectal cancer cell HCT116 images showed that the probe **19** is able to accumulate in lysosome effectively and a good overlap with LTG was observed (Fig. 24A–D). Furthermore, the HCT116 cells with O-(2-acetamido-2-deoxy-D-glucopyranosylidene) amino-N-phenylcarbamate (PUGNAc) showed significant decrease in bright red emission (Fig. 24E–H), thus, confirming Hex-catalyzed hydrolysis of probe **19**. Further, the utility of derivative **19** is also demonstrated for live imaging of Hex in animals.

Based on the above principle, Liu et al. reported a specific probe **21** for lysosomal β-galactosidase in ovarian cancer cells (SKOV-3 cells).[64] The standard MTT results showed 95% viability of SKOV-3 cells after incubation for 24 h at high concentration of **21** (100 μM).

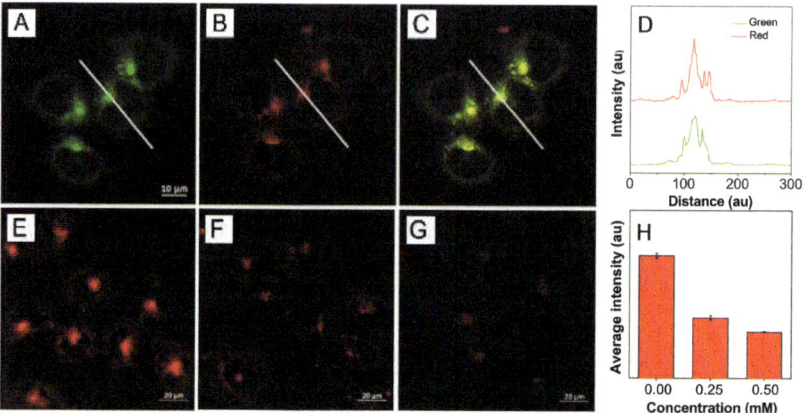

Fig. 24 (A) Confocal image of LTG DND-26 (B) Image of **19** (C) Merged images. (D) Intensity profiles. Images of cells stained with **19** in the absence (E) and presence of PUGNAc (F and G). (H) Average intensity of cells stained with **19** in the absence and presence of PUGNAc. *Copyright 2019, American Chemical Society.*

Further, lysosome targeting ability and the potential of probe **21** for lysosomal β-galactosidase detection was examined via staining SKOV-3 cells with probe **21** molecules and Lysosome Tracker **22**. The confocal microscopic images of Lysosome Tracker **22** (Fig. 25A) and LTR (Fig. 25B) showed lysosome-confined emission in SKOV-3 cells (Fig. 25C).

Pearson's correlation coefficient for Lysosome Tracker **22** and commercial LTR is calculated to be 0.94 (Fig. 25D), which indicate that Lysosome Tracker **22** possesses excellent lysosome-targeting ability. The SKOV-3 cells treated with both **21** and Lysosome Tracker **22** are "light up" and merge well in lysosomes (Fig. 25E–G) having Pearson's correlation coefficient of 0.83 (Fig. 25H). The "light up" is attributed to the removal of protected

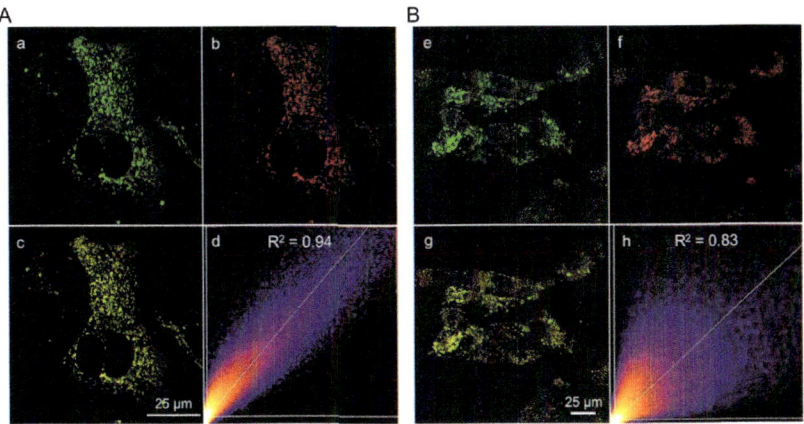

Fig. 25 (A) Confocal imaging to examine Lysosome-targeting ability of **22** in SKOV-3 cells. (a) Image of **22**; (b) LTR; (c) Merged images. (d) The co-localization analysis of **22** and **21**. (B) Verification of lysosome co-stain ability of Lysosome Tracker **22** and **21** in SKOV-3 cells. (e) Image of Lysosome Tracker **22**; (f) Image of **21**; (g) Merged image of (e) and (f). (h) Co-localization of Lysosome Tracker **22** and **21**. *Copyright 2020, American Chemical Society.*

group of hydroxyl moiety of **21** upon reaction with β-galactosidase in ovarian cancer cells to produce **23** (Scheme 6).

The good biocompatibility and specificity of **21** toward lysosomes make it suitable to visualize β-galactosidase in ovarian cancer cells.

4. Conclusions

This chapter gives an account of studies on development of AIE active materials for specific lysosome imaging. Thanks to the AIE phenomenon, aggregates of AIEgens show the bright emission even after accumulation in the lysosome. AIE phenomenon is further applied for development of "turn-on" probe for sensing of wide variety of bioanalytes such as H_2S, HClO, β-galactosidase, etc., in lysosome. The high photostability and biocompatibility of AIEgens are used for tracking the morphology, distribution and movements of lysosomes in living cells. Further, the reports regarding development of NIR emissive AIEgens with lysosome specific functionality having in vivo applications is summarized. The potential of red emissive AIEgens with multiphoton absorption is yet to be explored in advanced biomedical and energy applications. We expect advanced clinical applications of long wavelength emissive AIEgens in near future. We believe that this

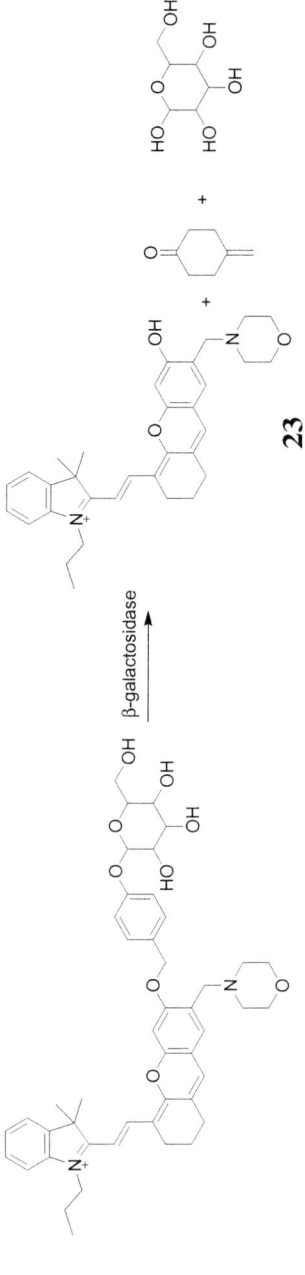

Scheme 6 Reaction of **21** with β-galactosidase.

chapter will create more interest and ideas for researchers to further develop AIE active probes for imaging lysosomes and monitoring their related activities.

Acknowledgment

We are thankful to UGC, CSIR, DRDO, DST and SERB for the financial assistance. We are thankful to Guru Nanak Dev University for research facilities. A.G is grateful to DAV University, Jalandhar for research facilities.

References

1. Appelmans F, Wattiaux R, De Duve C. Tissue fractionation studies. 5. The association of acid phosphatase with a special class of cytoplasmic granules in rat liver. *Biochem J.* 1955;59:438–445.
2. Cai Y, Gui C, Samedov K, et al. An acidic pH independent piperazine-TPE AIEgen as a unique bioprobe for lysosome tracing. *Chem Sci.* 2017;8:7593–7603.
3. Zhu H, Fan J, Du J, Peng X. Fluorescent probes for sensing and imaging within specific cellular organelles. *Acc Chem Res.* 2016;49:2115–2126.
4. de Duve C, Wattiaux R. Functions of lysosomes. *Annu Rev Physiol.* 1966;28:435–492.
5. Zhang W, Zhou F, Wang Z, Zhao Z, Qin A, Tang BZ. A photostable AIEgen for specific and real-time monitoring of lysosomal processes. *Chem Asian J.* 2019;10: 1662–1666.
6. Devany J, Chakraborty K, Krishnan Y. Subcellular nanorheology reveals lysosomal viscosity as a reporter for lysosomal storage diseases. *Nano Lett.* 2018;18:1351–1359.
7. Wang YF, Zhang T, Liang XJ. Aggregation-induced emission: Lighting up cells, revealing life. *Small.* 2016;12:6451–6477.
8. Hu F, Liu B. Organelle-specific bioprobes based on fluorogens with aggregation-induced emission (AIE) characteristics. *Org Biomol Chem.* 2016;14:9931–9944.
9. Fan F, Nie S, Yang D, Luo M, Shi H, Zhang YH. Labeling lysosomes and tracking lysosome-dependent apoptosis with a cell-permeable activity-based probe. *Bioconjug Chem.* 2012;23:1309–1317.
10. Satori CP, Henderson MM, Krautkramer EA, Kostal V, Distefano MM, Arriaga EA. Bioanalysis of eukaryotic organelles. *Chem Rev.* 2013;113:2733–2811.
11. Luker GD, Luker KE. Optical imaging: Current applications and future directions. *J Nucl Med.* 2007;49:1–4.
12. Weissleder R, Pittet MJ. Imaging in the era of molecular oncology. *Nature.* 2008; 452:580–589.
13. Zhang J, Cheng P, Pu K. Recent advances of molecular optical probes in imaging of β-galactosidase. *Bioconjug Chem.* 2019;30:2089–2101.
14. Fernández-Suárez M, Ting AY. Fluorescent probes for super-resolution imaging in living cells. *Nat Rev Mol Cell Biol.* 2008;9:929–943.
15. Domaille DW, Que EL, Chang CJ. Synthetic fluorescent sensors for studying the cell biology of metals. *Nat Chem Biol.* 2008;4:168–175.
16. An BK, Kwon SK, Jung SD, Park SY. Enhanced emission and its switching in fluorescent organic nanoparticles. *J Am Chem Soc.* 2002;124:14410–14415.
17. Michalet X, Pinaud F, Bentolila L, et al. Quantum dots for live cells, in vivo imaging, and diagnostics. *Science.* 2005;307:538–544.
18. Shaner NC, Steinbach PA, Tsien RY. A guide to choosing fluorescent proteins. *Nat Methods.* 2005;2:905–909.

19. Liang J, Li K, Liu B. Visual sensing with conjugated polyelectrolytes. *Chem Sci.* 2013;4:1377–1394.
20. Li J, Zhu JJ, Xu K. Fluorescent metal nanoclusters: From synthesis to applications. *TrAC Trends Anal Chem.* 2014;58:90–98.
21. Förster T, Kasper K. Ein Konzentrationsumschlag der Fluoreszenz des Pyrens. *Zeitschrift für Elektrochemie, Berichte der Bunsengesellschaft für physikalische Chemie.* 1955;59:976–980.
22. Luo J, Xie Z, Lam JW, et al. Aggregation-induced emission of 1-methyl-1,2,3,4,5-pentaphenylsilole. *Chem Commun.* 2001;1740–1741.
23. Ding D, Li K, Liu B. Tang BZ, bioprobes based on AIE fluorogens. *Acc Chem Res.* 2013;46:2441–2453.
24. Hong Y, Lam JW, Tang BZ. Aggregation-induced emission: Phenomenon, mechanism and applications. *Chem Commun.* 2009;4332–4353.
25. Mei J, Leung NL, Kwok RT, Lam JW, Tang BZ. Aggregation-induced emission: Together we shine, united we soar. *Chem Rev.* 2015;115:11718–11940.
26. Mei J, Hong Y, Lam JW, Qin A, Tang Y, Tang BZ. Aggregation-induced emission: The whole is more brilliant than the parts. *Adv Mater.* 2014;26:5429–5479.
27. Shi H, Liu J, Geng J, Tang BZ. Liu B, specific detection of integrin $\alpha_v\beta_3$ by light-up bioprobe with aggregation-induced emission characteristics. *J Am Chem Soc.* 2012;134:9569–9572.
28. Hong Y, Lam JW, Tang BZ. Aggregation-induced emission. *Chem Soc Rev.* 2011;40:5361–5388.
29. Leung CWT, Hong Y, Chen S, Zhao E, Lam JWY, Tang BZ. A. Photostable, AIE luminogen for specific mitochondrial imaging and tracking. *J Am Chem Soc.* 2013;135:62–65.
30. Li Y, Wu Y, Chang J, Chen M, Liu R, Li F. A bioprobe based on aggregation induced emission (AIE) for cell membrane tracking. *Chem Commun.* 2013;49:11335–11337.
31. Gao M, Hu Q, Feng G, Tang BZ, Liu B. A fluorescent light-up probe with "AIE + ESIPT" characteristics for specific detection of lysosomal esterase. *J Mater Chem B.* 2014;2:3438–3442.
32. Wang E, Zhao E, Hong Y, Lam JWY, Tang BZ. A highly selective AIE fluorogen for lipid droplet imaging in live cells and green algae. *J Mater Chem B.* 2014;2:2013–2019.
33. Hu QL, Gao M, Feng GX, Liu B. Mitochondria-targeted cancer therapy using a light-up probe with aggregation-induced-emission characteristics. *Angew Chem, Int Ed.* 2014;53:14225–14229.
34. Hu F, Huang Y, Zhang G, Zhao R, Yang H, Zhang D. Targeted bioimaging and photodynamic therapy of cancer cells with an activatable red fluorescent bioprobe. *Anal Chem.* 2014;86:7987–7995.
35. Yuan Y, Zhang CJ, Gao M, Zhang R, Tang BZ, Liu B. Specific light-up bioprobe with aggregation-induced emission and activatable photoactivity for the targeted and image-guided photodynamic ablation of cancer cells. *Angew Chem, Int Ed.* 2015;54:1780–1786.
36. Fu W, Yan C, Guo Z, et al. Rational design of near-infrared aggregation-induced-emission-active probes: In situ mapping of amyloid-β plaques with ultrasensitivity and high-fidelity. *J Am Chem Soc.* 2019;141:3171–3177.
37. Zhang J, Wang Q, Guo Z, et al. High-fidelity trapping of spatial–temporal mitochondria with rational design of aggregation-induced emission probes. *Adv Funct Mater.* 2019;29:1808153.
38. Wang Y, Zhang Y, Wang J, Liang XJ. Aggregation-induced emission (AIE) fluorophores as imaging tools to trace the biological fate of nano-based drug delivery systems. *Adv Drug Deliv Rev.* 2019;143:161–176.
39. Wang L, Xiao Y, Tian W, Deng L. Activatable rotor for quantifying lysosomal viscosity in living cells. *J Am Chem Soc.* 2013;135:2903–2906.

40. Liu T, Liu X, Spring DR, Qian X, Cui J, Xu Z. Quantitatively mapping cellular viscosity with detailed organelle information via a designed pet fluorescent probe. *Sci Rep*. 2014;4:5418.
41. Wu L, Li X, Huang C, Jia N. Dual-modal colorimetric/fluorescence molecular probe for ratiometric sensing of pH and its application. *Anal Chem*. 2016;88:8332–8338.
42. Guo B, Jing J, Nie L, et al. A lysosome targetable versatile fluorescent probe for imaging viscosity and peroxynitrite with different fluorescence signals in living cells. *J Mater Chem B*. 2018;6:580–585.
43. Li LL, Li K, Li MY, et al. Q. Bodipy-based two-photon fluorescent probe for real-time monitoring of lysosomal viscosity with fluorescence lifetime imaging microscopy. *Anal Chem*. 2018;90:5873–5878.
44. Hou L, Ning P, Feng Y, et al. Two-photon fluorescent probe for monitoring autophagy via fluorescence lifetime imaging. *Anal Chem*. 2018;90:7122–7126.
45. Li X, Zhao R, Wang Y, Huang C. A new GFP fluorophore-based probe for lysosomes labelling and tracing lysosomal viscosity in live cells. *J Mater Chem B*. 2018;6:6592–6598.
46. Chen W, Gao C, Liu X, et al. Engineering organelle-specific molecular viscosimeters using AIE luminogens for live cell imaging. *Anal Chem*. 2018;90(15):8736–8741.
47. Lou X, Zhang M, Zhao Z, et al. A photostable AIE fluorogen for lysosome-targetable imaging of living cells. *J Mater Chem B*. 2016;4:5412–5417.
48. Shi H, Kwok RT, Liu J, Xing B, Tang BZ, Liu B. Real-time monitoring of cell apoptosis and drug screening using fluorescent light-up probe with aggregation-induced emission characteristics. *J Am Chem Soc*. 2012;134:17972–17981.
49. Wang X, Liu H, Li J, et al. A fluorogenic probe with aggregation-induced emission characteristics for carboxylesterase assay through formation of supramolecular microfibers. *Chem-Asian J*. 2014;9:784–789.
50. Liang J, Kwok RT, Shi H, Tang BZ, Liu B. Fluorescent light-up probe with aggregation-induced emission characteristics for alkaline phosphatase sensing and activity study. *ACS Appl Mater Interfaces*. 2013;5:8784–8789.
51. Wai C, Leung T, Wang Z, et al. A lysosome-targeting AIEgen for autophagy visualization. *Adv Healthc Mater*. 2016;5:427–431.
52. Ouyang J, Zang Q, Chen W, et al. Bright and photostable fluorescent probe with aggregation-induced emission characteristics for specific lysosome imaging and tracking. *Talanta*. 2016;159:255–261.
53. Antunes F, Cadenas E, Brunk UT. Apoptosis induced by exposure to a low steady-state concentration of H2O2 is a consequence of lysosomal rupture. *Biochem J*. 2001;356:549–555.
54. Wu J, Li Y, Tan C, et al. Aggregation-induced near-infrared emitting platinum(II) terpyridyl complex: Cellular characterisation and lysosomespecific localization. *Chem Commun*. 2018;54:11144–11147.
55. Chen S, Wang H, Hong Y, Tang BZ. Fabrication of fluorescent nanoparticles based on AIE luminogens (AIE dots) and their applications in bioimaging. *Mater Horiz*. 2016;3:283–293.
56. Cheung NS, Peng ZF, Chen MJ, Moore PK, Whiteman M. Hydrogen sulfide induced neuronal death occurs via glutamate receptor and is associated with calpain activation and lysosomal rupture in mouse primary cortical neurons. *Neuropharmacology*. 2007;53:505–514.
57. Paul BD, Snyder SH. H_2S signalling through protein sulfhydration and beyond. *Nat Rev Mol Cell Biol*. 2012;13:499–507.
58. Zhang P, Hong Y, Wang H, et al. Selective visualization of endogenous hydrogen sulfide in lysosomes via using aggregation induced emission dots. *Polym Chem*. 2017;8:7271–7278.

59. Hong Y, Wang H, Xue M, et al. Rational design of ratiometric and lysosome-targetable AIE dots for imaging endogenous HClO in live cells. *Mater Chem Front.* 2019; 3:203–208.
60. Zhang W, Liu W, Li P, et al. Reversible two-photon fluorescent probe for imaging of hypochlorous acid in live cells and *in vivo*. *Chem Commun.* 2015;51:10150–10153.
61. Yuan A, Wu J, Tang X, Zhao L, Xu F, Hu Y. Application of near-infrared dyes for tumor imaging, photothermal, and photodynamic therapies. *J Pharm Sci.* 2013;102:6–28.
62. Zhuang J, Yang H, Li Y, Wang B, Li N, Zhao N. Efficient photosensitizers with aggregation-induced emission characteristics for lysosome- and gram-positive bacteria-targeted photodynamic therapy. *Chem Commun.* 2020;56:2630–2633.
63. Wang Q, Li C, Chen Q, et al. Lysosome-targeting red-emitting aggregation-induced emission probe with large stokes shift for light-up in situ visualization of β-N-acetylhexosaminidase. *Anal Chem.* 2019;91:12611–12614.
64. Li X, Pan Y, Chen H, et al. A specific near-infrared probe for ultrafast imaging of lysosomal β-galactosidase in ovarian cancer cells. *Anal Chem.* 2020;92:5772–5779.

CHAPTER SEVEN

Aggregation induced emission (AIE) materials for mitochondria imaging

Satish Deshmukh[a,†], Madan R. Biradar[b,c,†], Kiran Kharat[d,*], and Sidhanath Vishwanath Bhosale[b,c,*]

[a]Department of Chemistry, MSPMs' Deogiri College, Aurangabad, India
[b]Polymers and Functional Materials Division, CSIR-Indian Institute of Chemical Technology, Hyderabad, Telangana, India
[c]Academy of Scientific and Innovative Research (AcSIR), Ghaziabad, Uttar Pradesh, India
[d]KETs V.G. Vaze College, Mumbai, India
*Corresponding authors. e-mail address: krkharat@gmail.com; bhosale@iict.res.in

Contents

1. Introduction	180
2. Tetraphenylethene based AIEgens	181
2.1 TPE based probes for the detection of autophagy and/or mitophagy	182
2.2 TPE based chromophores for imaging of mitochondria in live cells	183
2.3 Molecules for photodynamic therapy	183
2.4 Mitochondrial membrane potential sensing molecules	185
2.5 AIEgens for the tumor tracking	187
2.6 Pyridinium modified TPE for DNA detection and mitochondria imaging	189
2.7 Dual functional AIE molecule	190
2.8 Mitochondrion-targeted AIE bio-probe for tumor cells	191
2.9 All-in-one molecular AIE theragnostic probe	191
3. TPA derivatives for the mitochondria detecting	192
3.1 Mitochondrion-targeted imaging and superoxide anion tracking in living cells	192
3.2 TPE derivatives based on D-π-A architecture	193
4. Miscellaneous AIEgens probe for mitochondria imaging	194
4.1 Cationic pyridinium group based salicyladazine fluorophore for selective mitochondrial targeting	194
4.2 Dual color imaging of cell membrane and mitochondria	195
4.3 Cyanostilbene skeleton as the AIE active core for the mitochondria targeting	195
4.4 AIEgen with organelle-specific emission for nucleolus and mitochondria imaging	197

[†] First two authors contributed equally.

4.5 Mitochondria-targeted polydopamine nanoparticle with AIE as photosensitizer   197
4.6 Spatial–temporal aggregation-induced emission probes for mitochondria tracing   198
4.7 Mitochondria-detected tumor therapy by using AIE active light-up probe   199
4.8 AIE-TADF based probe for mitochondrion tracing in living cells   199
5. Future prospectus   200
Acknowledgments   201
References   201

Abstract

Mitochondria are energy producing organelle of the eukaryotic cells. The main activities of mitochondria monitored by various marker molecules are autophagy detection, estimation of Reactive Oxygen Species (ROS), mitochondrial death and Photodynamic therapy in cancer cells. Due to the advantages of specificity and sensitivity, aggregation induced emission (AIE) is now popular for the mitochondria labeling. In this chapter, we would like to discuss three major types of AIEgens probe used in mitochondrial staining. There are three different types of AIEgens available for mitochondrial detection and sensing based on their different structural motifs. The first type of AIEgens is tetraphenylethene (TPE) based molecules. Due to simple engineering architecture, TPE based AIEgens are widely employed in bioimaging applications. AIEgen such as triphenylphosphine (TPP), and triphenylamine (TPA) are also employed as a novel building block. These are successfully used as exceptional lipid droplet (LD)-specific bio probes in cell imaging, assurance of cell combination, and photodynamic cancer cell removal. The third group is the miscellaneous AIEgens probe involved in mitochondria imaging.

1. Introduction

In 2001, Tang and coworkers introduced the term aggregation–induced emission[1] AIE luminogens (AIEgens) are found non–emissive in solution just because of non–radioactive decay. Aggregation caused quenching (ACQ) fluorophores are stronger than the AIEgens. These weak AIEgens form nano-aggregates which produces strong fluorescence.[1,2] AIEgens are used for the staining and detection of live cells, fixed cells and bio–imaging, diagnosis and therapy.[2,3]

Mitochondria are mainly involved in the apoptosis (programmable cell death in multicellular organism) and cell signaling process.[4–7] Its morphology is directed by a bunch of proteins and change of these proteins is accounted for neurodegenerative and cardiovascular diseases.[5–7] The significant capacity of mitochondria is to create energy and apoptosis induction.[8,9] The role of mitochondria in the induction of apoptosis has appeared as a new field of drug discovery in cancer cell biology research. To induce the death of the cell by apoptosis, the drug molecule must enter the mitochondria and induce the irreversible type of apoptosis in malignant cells.[9,10] Mitochondria are double membrane organelle having outer layer and inner layer separated by an intermembrane space. Due to increased metabolic activity in the cancer cells, a greater number of mitochondria with greater mitochondrial membrane potential (MMP) were reported to produce ATP. The greater membrane potential in mitochondrial membrane was the reason for designing mitochondria-targeting probes.[10,11]

Mitochondrial structure and function are analyzed by various chromogenic agents, biochemical markers, fluorescent markers and gene expression analysis assays.[12] In this chapter, we would like to discuss three major types of AIEgens used in mitochondrial staining. All these three major types of AIEgens are available for the mitochondrial detection and sensing based on their different structural motifs. The first type of AIEgens is TPE based molecules and have been widely employed for different bioimaging applications because of their simpler engineering architecture.[13,14] In the following section in this chapter, we discussed AIEgens such as triphenylphosphine (TPP), triphenylamine (TPA) and thiophene as building blocks. The third group is miscellaneous AIEgens Probe involved in the mitochondria imaging and selective cell imaging organelles.[15,16]

2. Tetraphenylethene based AIEgens

The tetraphenylethene (TPE) based AIEgens are propeller shaped popular molecules. Very recently, it has been confirmed that in aggregate state AIEgens have ability to execute ROS production.[14] These exceptional properties are ideal for mitochondria-focused on photodynamic treatment (PDT) as the accretion of AIEgens in mitochondria or encapsulation of AIEgens by mitochondria-based nanocarriers won't adjust their fluorescence and ROS generation.[14] Some probes are developed to study mitochondrial morphology with the different processes in living adipose cells.[17] TPE and its derivatives appeared as a promising molecule for bioimaging of organelles in cells due to its simple design and variety of modifications. The TPP molecules are the target molecules for the TPE derivatives. TPP is a functional group used to enter the molecular probes into mitochondria by its ability to dissolve in fats (lipophilicity) and electrophoretic effect (migration of charged particles under electric field).

2.1 TPE based probes for the detection of autophagy and/or mitophagy

In autophagy and mitophagy process the major targets are mitochondria. Zhang et al. synthesized isothiocyanate functionalized TPE-Py-NCS **(1)** having yellow emission with superior photostability was employed for targeting mitochondria in live cells (Fig. 1A). Fluorescence images show red and yellow fluorescence spots observed in the lysosome and mitochondria region. After 1.5 min, a new red spot observed Fig. 1B, which shows the formation of acidic autophogosome to start the mitophagy process.[18]

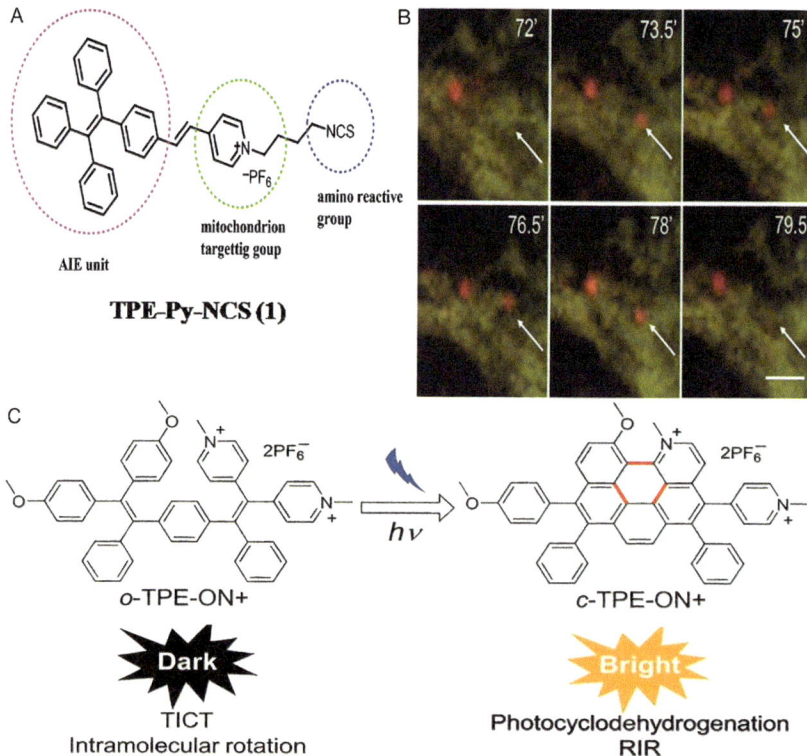

Fig. 1 (A) Probe TPE-Py-NCS **(1)**, (B) confocal images of HeLa cells treated by probe **(1)**. Copyright @ 2015, Royal Society of Chemistry, and (C) mechanism of photocyclodehydrogenation of O-TPE-ON$^+$. Panel (B): Reproduced with permission from Zhang W, Kwok RTK, Chen Y, et al. Real-time monitoring of the mitophagy process by a photostable fluorescent mitochondrion-specific bioprobe with AIE characteristics. Chem Commun. 2015;51:9022–9025. Panel (C): Reproduced with permission from Gu X, Zhao E, Zhao T, et al. Mitochondrion-specific photoactivatable fluorescence turn-on AIE-based bioprobe for localization super-resolution microscope. Adv Mater. 2016;28:5064–5071. Copyright @ 2016, Wiley-VCH.

2.2 TPE based chromophores for imaging of mitochondria in live cells

Mitochondria are called as the power house of cells. The monitoring of the mitochondria in live cell helps in the analysis of cell health. The AIEgen molecules are reported for the detection of the mitochondrial activities in the cell metabolism and the cell cycle.

2.2.1 Bioprobe based molecules

The imaging of mitochondria morphology and dynamics in live cells with higher resolution is difficult by using traditional fluorescence microscopy and organic dyes.[19] To overcome this difficulty and to detect mitochondria efficiently new chromophore named as o-TPE-ON$^+$ has been successfully established (Fig. 1C). In aqueous solution, o-TPE-ON$^+$ is practically non-emissive.[20]

Under UV illumination, with high emission productivity fluorophore c-TPE-ON$^+$ was generated *via* photoactivation of o-TPE-ON$^+$. Moreover, excitation at 561 nm with strong laser beam c-TPE-ON$^+$ resulted into spontaneous blinking and was photobleached. These characteristics make the chromophore an efficient entity for stochastic optical reconstruction microscopic imaging (STORM) of mitochondria in living cells.[21–23]

2.2.2 A heteroatom-containing luminogenic living cells with high photostability

Zhao *et al.* reported heteroatom-containing yellow-emissive probe TPE-Py (2) bearing a pyridinium moiety *via* modification of the vinyl subunit in the chromophore. In the solution state probe (2) is very less emissive whereas in poor solvent and/or in solid state upon aggregation it emits strongly. This makes probe (2) as an attractive entity to detect mitochondria in live cells with great photostability (Fig. 2C).[23] During solid-state emission of probe (2) shows reversibly switching between green and yellow color by means of grinding-heating and grinding- fuming as the physical methods (Fig. 2B). It shows high conversion from the amorphous to the crystalline state and *vice-versa*.

2.3 Molecules for photodynamic therapy

Lately, photodynamic therapy (PDT) has drawn a lot of consideration as a noninvasive and safe malignant growth treatment technique because of its fine controllability, great selectivity, low fundamental poisonousness, and negligible medication opposition interestingly to the traditional methods

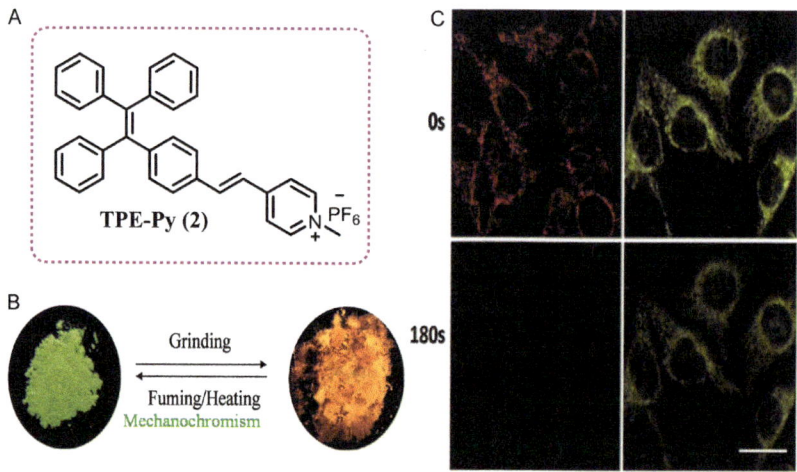

Fig. 2 (A) Probe TPE-Py structure (2); (B) study of mechanochromism effect; and (C) upon excitation at 405 and 560 nm wavelength UV light for 0 and 180 s, fluorescence imaging of HeLa cells treated with probe (2) with mitotracker red. *Reproduced with the permission from Zhao N, Li M, Yan Y, et al. Tetraphenylethene-substituted pyridinium salt with multiple functionalities: synthesis, stimuli-responsive emission, optical waveguide and specific mitochondrion imaging. J Mater Chem C 2013;1:4640–4646. Copyright @ 2013, Royal Society of Chemistry.*

such as chemo therapy radiotherapy. The photosensitizer molecules for killing the cells with oxygen are used in photodynamic therapy. This section is reporting few promising molecules with dual functionality.

2.3.1 AIE probe for mitochondria-targeting and photodynamic cancer therapy

Mitochondria are a primary energy supplier to cellular cells. PDT has taken more attention during the last few decades, due to its advantages over chemotherapy. Yang *et al.* designed an AIE-based two probes such as TEP-QN (3) and TPE-DQN (4) as shown in Fig. 3. In probe (4) there is the insertion of vinyl group in molecular backbone to increase the conjugation, induces emission at 650 nm. Both probes have cationic and AIE effects, which are used in mitochondria detecting and differentiating cancer cells over normal cells during the staining process.[24] These probes generate singlet oxygen during white light irradiation; the efficiency singlet oxygen generation of probe (4) has higher than probe (3). The cationic luminogens such as probes (3) and (4) with AIE properties are useful for mitochondria targeted photodynamic therapy. They have shown ~200 nm Stokes's shift, excellent biocompatibility and better photostability. It is reported that these can be

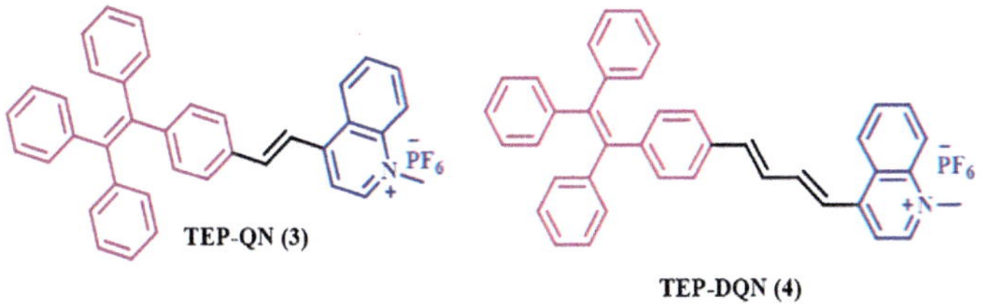

Fig. 3 Molecular structures of probes **(3)** and **(4)**.

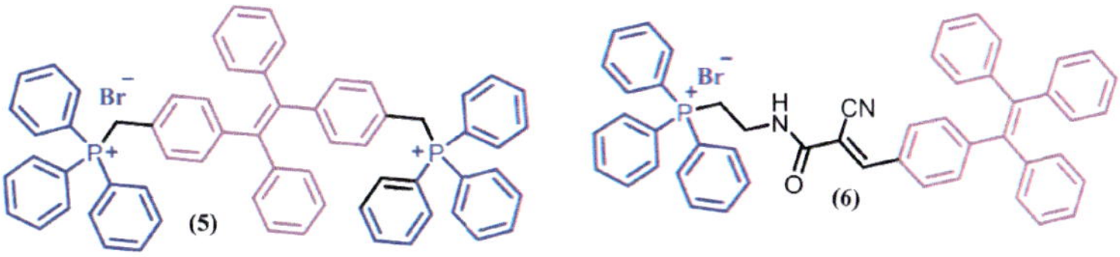

Fig. 4 Molecular structures of probes **(5)** and **(6)**.

used to stain the mitochondria specifically in malignant cells. The probe **(4)** was efficiently employed for selective photodynamic cancer cells ablation and image-guided PDT for tumor inhibition.

2.4 Mitochondrial membrane potential sensing molecules

During oxidative phosphorylation, the proton pumps generate mitochondrial membrane potential to keep the cell relatively stable. The normal physiological activities are also associated with the mitochondrial membrane potential.

2.4.1 TPE-TPP based probes to check mitochondrial membrane potential changes

As of late, Tang and his colleagues built up the AIE luminogen probes **(5)** and **(6)** (Fig. 4) which contain two triphenylphosphonium (TPP) subunits, which can be utilized mitochondrial imaging and tracking.[25] Also, Tang and co-workers displayed that probe **(5)** goes through exceptionally explicit binding to mitochondria in live cells. They observed that probe **(5)** exhibits good photostability and huge resistance than ecological changes. Authors treated live cells with probe **(5)** in presence of carbonyl cyanide *m*-chlorophenylhydrazone (CCCP), cells don't undergo an eminent change in their particular mitochondrial imaging and tracking.[25] They demonstrated that TPP moiety exhibits high lipophilicity, electrophoretic force, and good mitochondrial detecting ability. Tang proposed that the availability of two TPP groups in probe **(5)** have strong electrostatic repulsive interactions between them. This proposal assumes, incorporating only one TPP group to AIEgen *via* conjugation resulting in a competent mitochondrial membrane potential indicator.

2.4.2 Mitochondria targeted near-infrared probes

Zhao *et al.* developed TPE-Xan-In **(7)** chromophore for accurate and lifelong mitochondrial morphology imaging and as illustrated in Fig. 5A. It is near-infrared (NIR) fluorescent probe that detects the mitochondria. Probe **(7)** shows absorption maxima at $\lambda_{max} = 660$ nm and at 743 nm fluorescence emission peak, which is in the NIR region. At neutral pH, probe **(7)** also displayed AIE properties for the mitochondria imaging with excellent photostability, awesome biocompatibility, and high resistance to potential changes across mitochondrial membrane. Probe **(7)** displayed a pH-responsive emission changes within pH 4.0–7.0 range due to the presence of $^-$OH functional group in xanthene molecular architecture, which showed its potential in monitoring pH and morphology effect of mitochondria in the biological research studies. The confocal microscopic images of probe **(7)** were compared with commercial available mitochondria tracker as displayed in Fig. 5B. There is excellent emission observed at high concentration of probe (7) compared to MTG.[26]

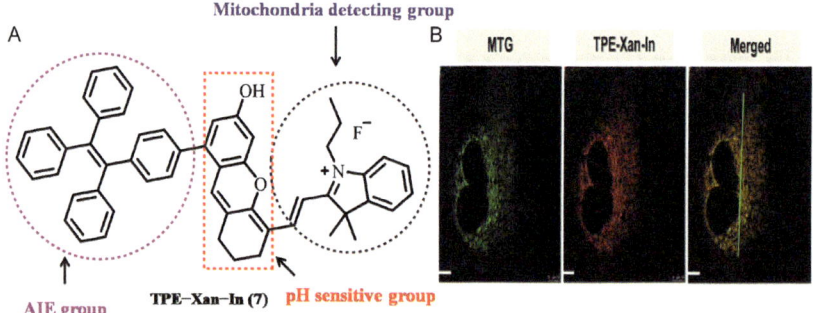

Fig. 5 (A) Structure of probe TPE-Xan-In **(7)** and (B) confocal laser scanning microscopy images for HeLa cells trained by probe **(7)** at neutral pH. *Reproduced with the permission from Zhao X, Chen Y, Niu G, Gu D, Wang J, Cao Y, Yin Y, Li X, Ding D, Xi R, Meng M. Photostable PH-sensitive near-infrared aggregation-induced emission luminogen for long-term mitochondrial tracking.* ACS Appl Mater Interfaces *2019;11:13134–13139. Copyright @ 2019, American Chemical Society.*

2.4.3 Mitochondria-targeted AIE probe for membrane potential and mouse sperm activity

Tang *et al.* reported two AIE-actives TPE-In **(8)** and TPE-Ph-In **(9)** (as shown in Fig. 6) compounds were synthesized by the Knoevenagel condensation. Luminogens of **(8)** and **(9)** emitting fluorescence in the red region. Compounds **(8)** shows fluorescence at $\lambda_{max} = 694$ nm in DMSO, while in the H_2O fraction has outstanding red fluorescence. The AIE effect is found more in **(8)** than the **(9)**. Upon aggregation of probe, the emission of **(9)** is increased about 70 times. Mostly, probe **(9)** exhibits more AIE effect and lesser cytotoxicity than **(8)**. Compound **(9)** was reported to stain mitochondria in living cells with excellent photostability.[27]

For the production of adenosine triphosphates (ATP), mitochondria oxidizes substrates as well as balance the proton gradient across the mitochondrial layer in the respiratory electron transport chain with huge membrane potential (DC_m). Mitochondrial membrane potential is deciding factor for the mitochondrial role in cell death, repair and regeneration.[27] Therefore an efficient method for DC_m and mitochondrial structure is needed in biomedical applications and recent diagnosis of various diseases. Fluorophore probe undergo intramolecular charge transfer (ICT) based on a D-A design which was employed as red emitters.[28–30] Indolium salt acts as electron acceptor subunit in the ICT luminogens played as an essential role for mitochondrial targeting probes.[31]

2.5 AIEgens for the tumor tracking
2.5.1 Mechanochromism based ionic AIE-gens with mitochondria imaging

The mechanochromism (the organic compound shows change in color after getting stress by mechanical grinding, crushing and milling) active AIEgens shows high fluorescence. Organic compounds having ionic species have been used to design the AIEgens, specifically TCPy and TCPyP **(10)** (Fig. 7A), tetraphenylatin, and phenyl acrylonitrile as important chromophores. Yang *et al.*

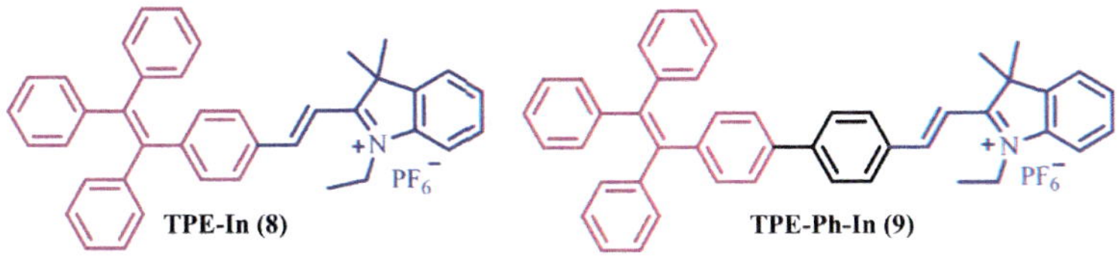

Fig. 6 Molecular structures of probes **(8)** and **(9)**.

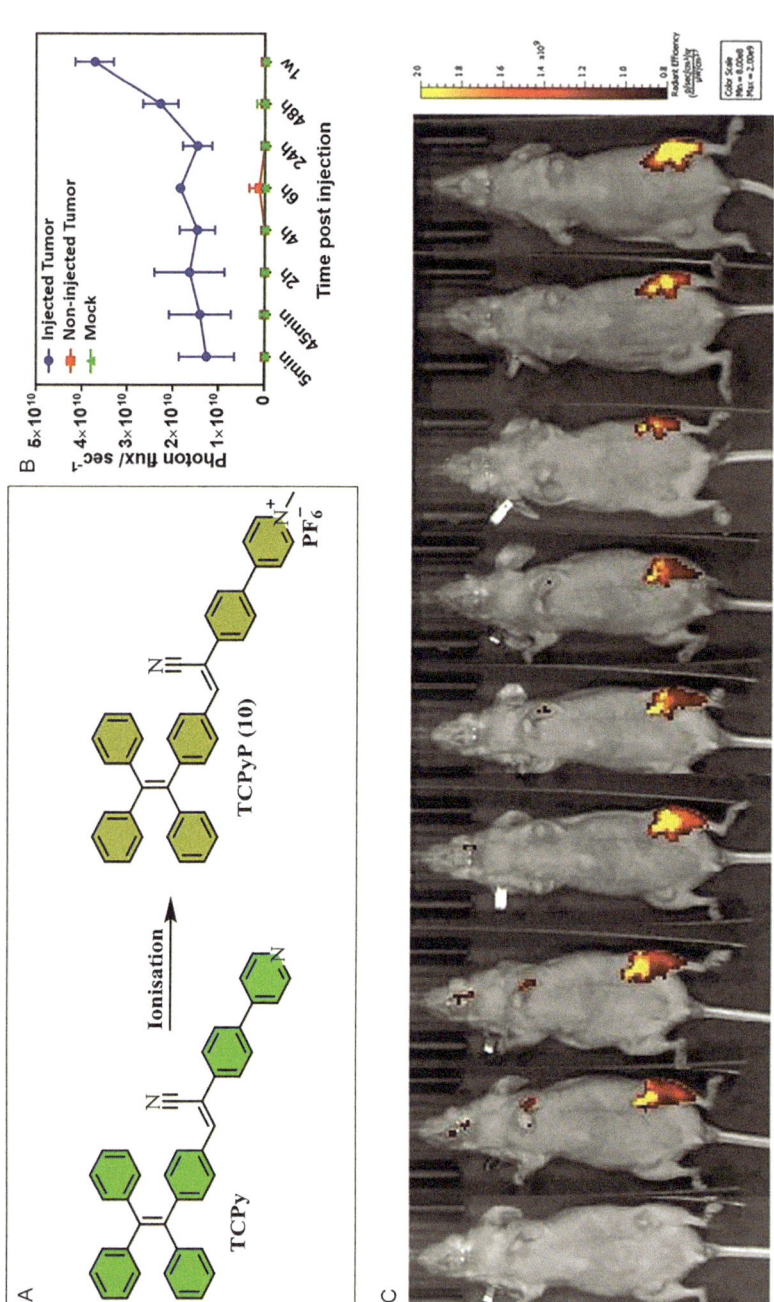

Fig. 7 (A) Molecular structure of TcPyP (**10**), (B) *in vivo* imaging of biodistribution of probe (**10**), and (C) luminescence intensity recorded at different time. *Reproduced with permission from ref. Yang X, Wang X, Wang Q, Hu P, et al. Achieving remarkable and reversible mechanochromism from a bright ionic AIEgen with high specificity for mitochondrial imaging and secondary aggregation emission enhancement for long-term tracking of tumors. Mater Chem Front. 2020;4:941–949. Copyright @ 2020, Royal Society of Chemistry.*

synthesized of TCPy and probe **(10)** using phenylacrylonitrile and TPE molecular building blocks, followed by insertion of pyridine to enhance the ionization position. The condensation reaction between 4-(1,2,2-triphenylvinyl) benzaldehyde and 2-(4-bromophenyl) acetonitrile in the presence of alkali as a catalyst and followed by Pd-catalyzed Suzuki coupling reaction with 4-(4,4,5,5-tetramethyl-1,3,2-dioxaborolan-2-yl) pyridine to form TCPy **(10)**. Later methylation on pyridine compounds and ion exchange for the synthesis of hexafluorophosphate from iodine were successfully conducted to offer the final product as probe **(10)** in Fig. 7A.[32]

Probe **(10)** has shown maximum illumination at 365 nm UV light in water/THF in 99% of underwater fractions concentration of 10 μM. Along with their AIE features, it shows high fluorescence quantum production yields. Two compounds were found suitable for *in vitro* bio imaging biological applications. In A549 cancer cells, dose-based cytotoxicity deciphers that showed no clear toxicity levels even though cells were treated with TCPy or probe **(10)** at concentrations up to 20 μm. The probe **(10)** can clearly visualize reticular mitochondria in the cells as shown in Fig. 7B. The luminescence intensity was plotted with respect to time for injected tumor and non-injected tumor. (Fig. 7C), after 1 week the intensity of injected tumor is very high compared to non-injected tumor. It showed intense orange emission obtained through visible light radiation at $\lambda_{max} = 405$ nm which also overlaps the pseudo-red emission of MTG with Pearson's coefficient showing a high accuracy of mitochondria-staining in living cells.[32]

2.6 Pyridinium modified TPE for DNA detection and mitochondria imaging

The cationic AIEgens can detect DNA through fluorescent change. DNA itself has anionic polyelectrolyte. The electrostatic interaction between cationic AIEgens and anionic DNA execute fluorescence. AIEgens have played an important role in live cell fluorescent imaging. The mitochondria have negative electrostatic potential in living cells and attract cationic. In the literature report cationic AIE-gens conjugated with phosphonium, pyridinium, quaternary ammonium and triphenylamine were employed for mitochondria specific fluorescent probes.[33–35] Wang *et al.* reported TPEDEPy-DBz **(11)** (Fig. 8) pyridinium modified TPE utilized for mitochondria imaging as well as DNA detection. The compound **(11)** exhibits excellent AIE properties. It is soluble in highly polar solvent and emission intensity faction 90% at the wavelength ($\lambda = 570$ nm). Similarly, the

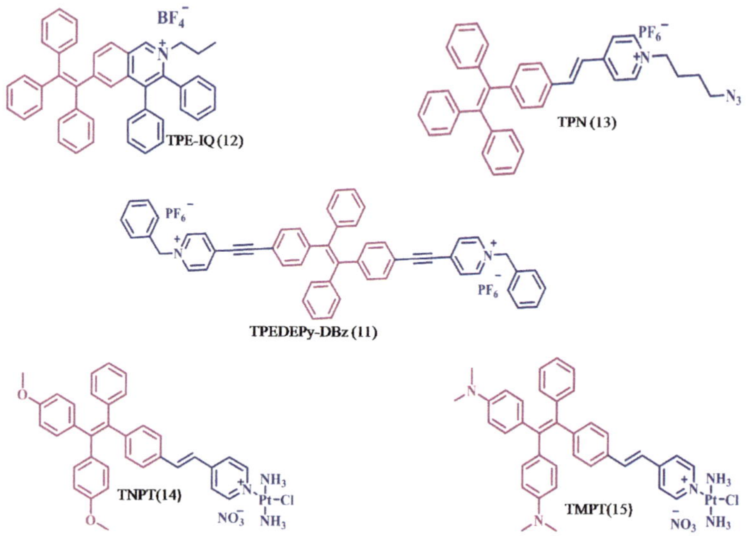

Fig. 8 Molecular structures of probes **(11)**–**(15)**.

UV–Vis spectra display the different range of solvents in λ abs of AIE-gen of **(11)** shifts 404 nm. Based on the large Stokes's shift and the experimental results suggest it is a promising cell imaging agent.[35]

2.7 Dual functional AIE molecule

Fluorophores having aggregation enhanced emission (AEE) property showed a dual role in mitochondria imaging and acts as a photosensitizer to generate ROS in mitochondria region to produce cell apoptosis. Recently Zhao and co-workers synthesized dual functional groups isoquinolinium-based molecule, designed with an AIE-active core TPEIQ **(12)** as shown in (Fig. 8). This fluorescence probe was used for mitochondrial detecting in live cells and worked photosensitizer for PDT.

Compound **(12)** shows AEE characteristics. The DMSO solution of it shows moderate fluorescence emission at 502 nm whereas in H_2O system leads to the aggregate formation, which shows enhanced emission of probe **(12)**.[36] It has the power to detect the mitochondria in living and fixed cells with low concentration of compound **(12)**. It can be used as a photosensitizer, compound **(12)**. Generate ROS upon UV irradiation and produces the cell death pathway—apoptosis effectively.[37,38] This promising molecule is having mitochondrial targeting ability with short incubation and irradiation time at low working concentration of the probe.

2.8 Mitochondrion-targeted AIE bio-probe for tumor cells

Zheng et al. reported that the bio-probe compound of TPN **(13)** (Fig. 8) with AIE, which can differentiate tumor cell diversity from blood leukocytes based on cell mitochondria differentiation.[39] The bio-probe compound **(13)** is a cell-permeant live-cell stain; it has an inefficient on cell feasibility. The compound **(13)** is made up of AIE unit, TPE, and a pyridinium group. It shows absorption at 405 nm in the polar solvent DMSO and it has an emission peak of 636 nm. Its AIE properties produce outstanding fluorescence "turn-on" after binding to the mitochondria just because leukocytes cells are the most abundant.[39]

TPN is a mitochondrion staining AIE bio-probe used in the identification of tumor cells in a normal tissue environment. Tumor cells stained with compound **(13)** (Fig. 8) emitted higher fluorescence than normal human leukocyte cells. This is because of a more prominent number of mitochondria with e raised mitochondrial membrane potential found in quickly developing tumor cells.[39]

The utilization of compound **(13)** is the antibody-free, live-cell labeling examines with low-cost, and has unimportant effect on cell suitability and uprightness, which permits high-quality investigation of tumor cell on single-cell level.

2.9 All-in-one molecular AIE theragnostic probe

Among the different cancer treatments, chemotherapy is one of the lives threatening treatments, with severe side effects. Similarly, PDT is also one of the cancer treatments which have good controllability, very small MDR and very few side effects. To overcome these problems, researchers developed new terminology theragnostic combination of Chemotherapy and PDT to kill a wide range of cancer cells. Prof. Bin Liu and co-workers reported AIE active fluorescence probes such as TNPT **(14)** (Fig. 8), TMPT **(15)** (Fig. 8) for the detection of mitochondria targeted chemo and photodynamic cancer cells.[40]

Photosensitizers with the keen combination of AIE subunit and cisplatin are effortlessly prepared for synergetic anticancer treatment. Cisplatin unit and similarly both hydrophilic-hydrophobic property, D-A strength of the recently designed AIE photosensitizer compound **(14)** shows great cell take-up with predominant mitochondria location of malignant cells, high chemotherapeutic viability, like cisplatin and ROS generation ability great. For compound **(15)** exhibits absorption maxima at 329 nm and which reached beyond 560 nm, while its fluorescence emission is 686 nm with a 266 nm Stokes shifts. The absorption maxima at 310 nm is ascribed to the normal π-π* transition of the modified TPE. The fluorescence quantum yields of **(14)** and **(15)** in H_2O were estimated to be 9.2% and 0.7%, which are lower than their precursors including TN (16.0%) and TM (2.1%), respectively.[40] Probe **(14)** has all in one feature, for example, good water-solubility, brilliant cell take-up, high mitochondria imaging, and intense chemotherapy like free cisplatin and strong PDT viability on malignant cells, while displaying fundamentally inhibited side effects on ordinary cells. Likewise, the cell cycle measure and apoptotic examination uncovered that the therapeutic mechanism of **(14)** is to fundamentally inhibit DNA replication and cause cell apoptosis. This investigation calls attention to new bits of knowledge and bearings for the future advancement of exceptionally viable double activity of molecular cancer drugs.[40]

3. TPA derivatives for the mitochondria detecting

From earlier literature, typical fluorescent probes incorporated triphenylphosphonium (TPP),[41,42] positively charged pyridine,[43] and quinolone derivatives[44] as the connecting sites with mitochondrion membranes having been disclosed. However, some great challenges have still existed, such as low selectivity and sensitivity, less spatial resolution, and invasive operation. Above all, some of them are mitochondrion-targeted, but are nonselective and sensitive toward ROS-associated dynamic fluctuation.

3.1 Mitochondrion-targeted imaging and superoxide anion tracking in living cells

The mitochondrion is the source of energy for the cells, which played an important role in energy generation. In the energy generation process electron transport involved during the oxidation-phosphorylation pathway, this is associated with ROS process. It mainly includes oxygen contain species such as singlet oxygen ($^1O_2$), superoxide anion (O_2^-), Hydrogen peroxide (H_2O_2), and hydroxyl radicals (^-OH). In which superoxide anion is primary ROS in living cells.

Recently, Ma et al. reported three TPA-Pyr based AIE molecules. They showed wide applications with AIE properties as well as cell-imaging applications. They have synthesized probes having the formula TPA-Pyr-Octane **(16)** (Fig. 9), TPA-Pyr-Br **(17)** (Fig. 9) and TPA-Pyr-Thiourea **(18)** (Fig. 9). This probe shows AIE properties depending on the specific

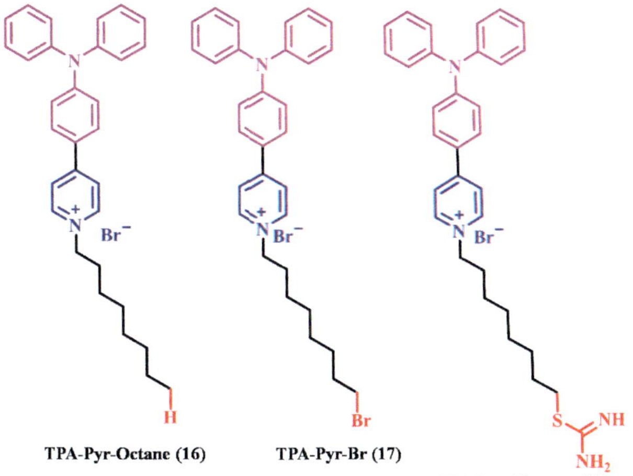

Fig. 9 Molecular structures of probes **(16)–(18)** based on TPA-Pyrmoity.

viscosity. In the system of glycerol and methanol at 9:1 (v/v) ratio, these probes emit fluorescence at 580 nm. Due to H-bonding between probes and mitochondria protein, it is helpful for detecting mitochondria imaging and for tracing of superoxide anion in living cells. The AIEgen with specificity to detect the ROS from mitochondria will play a significant role in real time detection of the mitochondrial mechanism and various diseases related to mitochondria.[45]

3.2 TPE derivatives based on D-π-A architecture

In this study, the triphenylamine group was utilized as the electron-donating cite and pyridinium group as the electron-accepting cite just as the mitochondria-focusing site. Two-photon AIEgens such as **TPPM (19)** and **TTPM (20)** dependent on a D-π-A framework revealed for the mitochondria targeted TPA (Fig. 10A and B). Both AIEgens show strong AIE property with the emission upgrade up to 290-folds. AIE properties of probe **(19)** and **(20)** in THF/Water mixture is displayed in Fig. 10A and B. These are having high particularity for the mitochondria in living cells. These molecules are created as a two-photon bio probes for the live cell bio-imaging with a puncturing depth up to 150 μm. Probe **(19)** and **(20)** have high PDT productivity in HeLa cells under white light treatment has been demonstrated. With huge two-photon absorption, explicit for mitochondria in live dividing cells. It has excellent PDT efficiency to living cells as well as outstanding photostability and biocompatibility of these novel AIEgens

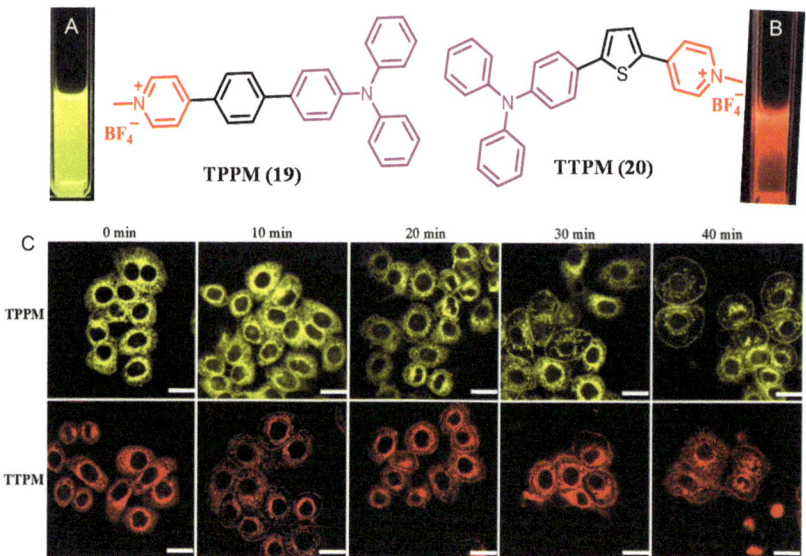

Fig. 10 (A and B) Photograph of probes **(19)** and **(20)**. The UV light irradiation at 365 nm in THF/water mixture solution (1:9 v/v). (C) Confocal images of probes **(19)** and **(20)** under white light irradiation. *Adopted and modified from Zhuang W, Yang L, Ma B, et al. Multifunctional two-photon AIE luminogens for highly mitochondria-specific bioimaging and efficient photodynamic therapy.* ACS Appl Mater Interfaces *2019;11:20715–20724.* Copyright @ 2019, American Chemical Society.

uncovers gigantic potential in clinical uses of two-photon cell and tissue bio-imaging and image-guided and mitochondria-focused on photodynamic malignancy therapy.[46]

4. Miscellaneous AIEgens probe for mitochondria imaging

4.1 Cationic pyridinium group based salicyladazine fluorophore for selective mitochondrial targeting

In the literature, a lot of fluorescent probes have been produced for mitochondrial targeting, for models such as rhodamines, rosamines, carbocyanines, styryl dyes, BODIPY and thiol-reactive mitotracker dyes. They have huge p-planar plans achieving little Stokes shifts (less than 40 nm), due to high-concentration accumulation in mitochondria their fluorescence would undergo through self-quenching.[47] To handle this task, a novel AIEgens such as AIE-MitoGreen-1 has been synthesized based on the salicyladazine fluorophore[48] and cationic pyridinium groups for selective mitochondrial imaging[48] this fluorophore utilized two novel emission steps: AIE through limitations of intramolecular rotation around the N—N bond and excited-state intramolecular proton transfer (ESIPT) through intramolecular hydrogen bonds.[49] A vital trait of brown adipose cells during separation is that their mitochondrial content and movement will increase to accommodate the greater metabolic requests. By seeing the movement in the number, morphology and sub-cellular limitations of mitochondria through fluorescence imaging, a non-invasive technique to find the separation phases of brown adipose cells could be given.[50]

4.2 Dual color imaging of cell membrane and mitochondria

The confocal laser microscopic images reveal that probe TPNPDA-C2 (**21**) detects the membrane and mitochondria of the cells through exciting fluorescence channels. As shown in Fig. 11. This probe has the ability to detect sudden changes in organelles during cell apoptosis and necrosis generated by ROS and cytotoxins. This dual colored probe has a lot of applications in cellular biology and disease mechanism.[51]

4.3 Cyanostilbene skeleton as the AIE active core for the mitochondria targeting

Many fluorescence probes are developed to detect mitochondrion changes such as pH, thickness and polarity by changing the microenvironment. AIEgens works on electrostatic attraction which may miss to detect mitochondria during mitophagy with decreased membrane potential which prompts erroneous thickness measurement.[52] So Wang et al. developed mitochondria-immobilized NIR-AIE active probe **CS-Py-BC(22)** which takes into an off-on fluorescence response to membrane thickness, consequently the real-time observing variation in membrane thickness during mitophagy (Fig. 12). This framework comprises of a cyanostilbene skeleton is not only an AIE active base but also viscosity-sensitive unit; this positively charged pyridinium group is used for the mitochondria-detecting and benzyl chloride subunit was used for the mitochondrial immobilization (Fig. 12).

In this study the membrane thickness was increased from 0.903 cP (0% glycerol) to 965 cP (99% glycerol) of probe **(22)** with this the fluorescence intensity was increased by 92-fold at 650 nm, which causes the limitations to both rotation and twisted intramolecular charge transfer in higher membrane thickness system.

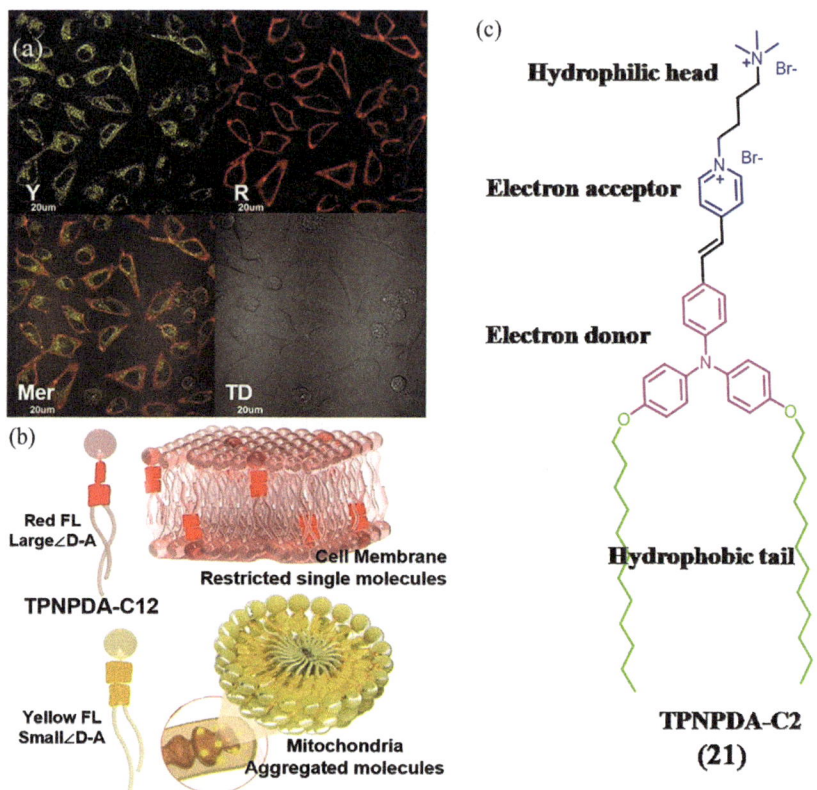

Fig. 11 (A) Confocal microscopic images of HeLa cell of probe **(21)**; (B) Schematic representation of simultaneous dual color imaging of cell membrane and mitochondria by probe **(21)**; and (C) molecular structure of TPNPDA-C2 probe **(21)**. *Adopted and modified from (a) Zheng Y, Ding Y, Ren J, Xiang Y, Shua Z, Tong A. Simultaneously and selectively imaging a cytoplasm membrane and mitochondria using a dual-colored aggregation-induced emission probe.* Anal Chem. *2020;92:14494–14500. (b) Wang X, Fan L, Wang S, et al. Real-time monitoring mitochondrial viscosity during mitophagy using a mitochondria-immobilized near-infrared aggregation-induced emission probe.* Anal Chem. *2021; 6:3241–3249. Copyright @ 2020, American Chemical Society.*

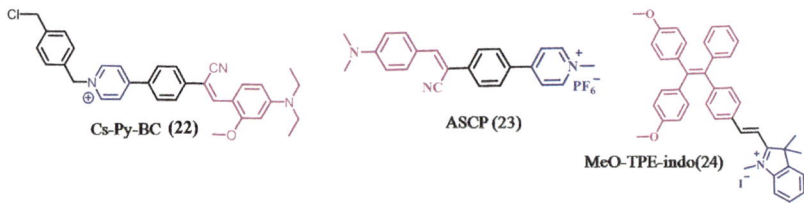

Fig. 12 Molecular structures of probes **(22)**–**(24)**.

4.4 AIEgen with organelle-specific emission for nucleolus and mitochondria imaging

Yu and co-workers developed AIE material based on α-cyanostilbene derivative, namely **ASCP (23)**, for simultaneously detecting both the organelles like nucleolus and mitochondria (Fig. 12). Compound **(23)** included N,N diethylamine group as an electron donating group to tune the emission of the red region; due to which emission take place at wavelength at 450 nm. The probe can light up nucleolus and mitochondria with higher sensitivity. Compound **(23)** has excellent optical properties due to high solubility in polar solvents and low solubility in aqueous solvents.[53] Due to the TICT effect Compound **(23)** emits strong orange light in dioxane solution, similarly in DMSO solution, it shows fluorescence property while the addition of toluene into DMSO solution. This solvent polarity played a vital role in the emission of light and transformation of color emission to orange.[54]

Compound **(23)** has different interactions with mitochondria in living cells and nucleolus which is observed in fluorescence spectroscopy. It has absolutely high brightness, outstanding biocompatibility and awesome photostability, due to these features it is a promising candidate for simultaneous mitochondria and nucleolus imaging.

4.5 Mitochondria-targeted polydopamine nanoparticle with AIE as photosensitizer

The therapeutic effect of PDT depends on photosensitizers and its photoefficiency depends on the selective wavelength of the source used as well as energy required to generate ROS. Recently nanocomposite materials are developed to deliver photosensitizers into cells and other organelles. Wang's group examined new probe for mitochondria targeting and for PDT as the potential for tumor treatment. The AIE active probe was designed based on D-π-A architecture. Here TPE as an AIEgens and shows superb photostability. OMe and indolium functional groups are attached on TPE fluorophore to form Me-O-TPE-indo **(24)** shown in Fig. 12. They have orchestrated nanocomposite with the combination of compound **(15)** and polydopamine nanoparticle (PDA) to form PMT in a nanocomposite. It can target mitochondria specifically by positively Indoliumion, and the D-π-A conjugate structure can create singlet oxygen effectively.[55] PDA has been used in cancer treatment as a transporter with phtotothermal therapy **(PTT)** work. Due to the presence of amino group on PDA nanoparticles, it executes non covalent infractions such as π–π interactions and hydrogen bonding between benzene rings and electrostatic interaction of cationic form of compound (24) and anionic form of PDA NPs. Reliable outcome as the carrying capacity of compound (24) comes up to 99.98%, and as its concentration increases with this carrying capacity diminishes hardly, keeping more than 84%.[56] To upgrade the therapeutic effectiveness and settling compound (24) in the aqueous system, PDA is used as a transporter to build PMT in a nanocomposite. The role of PMTi in fluorescent imaging and its time in tumor profile make it a superb candidate for surgical navigation in the coming years.

4.6 Spatial–temporal aggregation-induced emission probes for mitochondria tracing

We focused on the most proficient method to develop AIE-active probe for mitochondria targeting and image-guided therapy, outlining the connection between molecular aggregation state and organelle focusing capacity. Prof. Wei-Hong Zhu reported tricyano-methylene-pyridine (TCM), newer red-emitter included two TPP cation groups instead of the separated locations of AIE building architecture, focusing on mitochondria based on molecular TCM-1(25), TCM-2 (26) and TCM-3 (27) design probe (Fig. 13). Probe (25), EDG diethylamine alkoxybenzene can be productively upgraded in the NIR wavelength region for bio imaging and changing the position of positively charged TPP groups handle "off-on" properties and mitochondria targeting capacity. Additionally, substituted at *para-* and *ortho-*positions of TPP has good fluorescence "off-on" property, along these achieving the optimized mitochondrial. Due to the excellent AIE

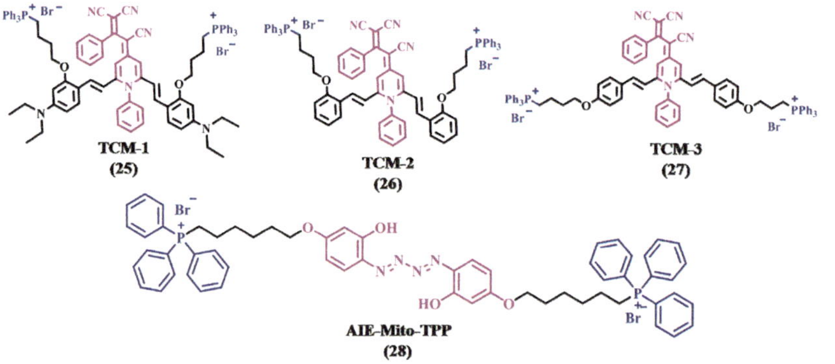

Fig. 13 Molecular structures of probes (25)–(28).

properties, probe **(25)** shows weak fluorescence in aqueous system whereas probe **(27)** in THF shows strong fluorescence. Thus TCM-1 **(25)** has proved a successful alternative to commercial mitochondrial trackers with commercial photography to achieve high reliability and temporal visualization.[57]

Probe **(25)** effectively kills cancer cells with altogether produced light-incited intracellular ROS by observing the energy gap between singlet and triplet. Therefore, we have strangely seen the AIE probe problem among the lipophilicity and hydrophilicity, disentangled the connections among the molecular aggregation state and organelle-focusing capacity. It proposes us savvy control in the fields of developing excellent probe to detect sub-cellular organelles for biomedicine, and for changing the commercially available MTG or MTR.[57]

4.7 Mitochondria-detected tumor therapy by using AIE active light-up probe

Fluorescence probe such as AIE-mito-TPP **(28)** (Fig. 13) was synthesized with TPP as core by Liu *et al*. The fluorescence of this probe must be turned on when there is a restriction of the rotation in nitrogen double bond and formation of the intramolecular hydrogen bond between them. This probe has outstanding ability to cell-culture media and indicates high cellular take up.[58,59] In the culture medium probe **(28)** have non-fluorescent properties, which selectively and immediately detect mitochondria in tumor cells compared to normal cells. Because of the amassing of probe **(28)** in tumor-cell causes decreasing the mitochondrial membrane thickness, formation of singlet oxygen in ROS process, and affect cancer-cell growth.[58]

4.8 AIE-TADF based probe for mitochondrion tracing in living cells

TADF materials are luminescent materials having delayed fluorescence with lifetime between nanosecond to milliseconds. Prof. C. Yang and co-workers reported the **NID-TPP (29)** probe in the mitochondrial matrix (Fig. 14). Probe **(29)** confirmed conglomeration upgraded TADF characteristics and orange-red emission; it is the best designed probes for time-resolved luminescence imaging (TRLI).NID and TPP$^+$ don't influence luminescence properties specifically for TADF emissions. (Fig. 14) The photoluminescence quantum yield of the Probe **(29)** in aqueous solution was 0.015% balanced with quinine sulfate in 0.1 M H_2SO_4. Because the strong

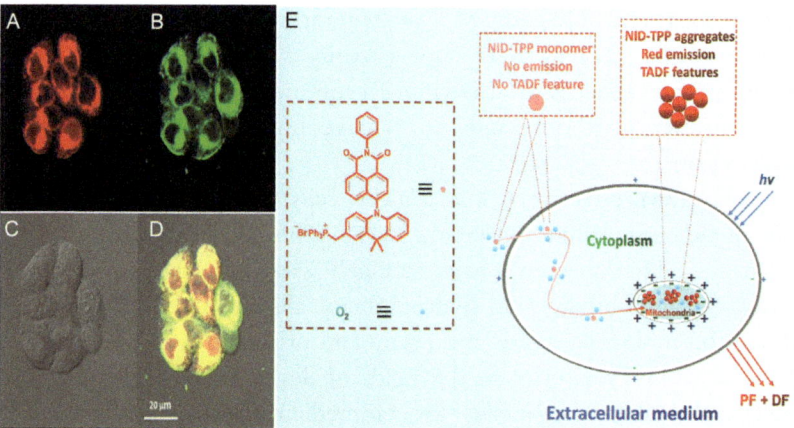

Fig. 14 (A) Luminescent image of probe **(29)**; (B) luminescent image mitotracker; (C) bright filled images; (D) merged images of probe **(29)** bright filled; and (E) mechanism of probe **(29)** for TRLI images. *Adopted and modified from Ni F, Zhu Z, Tong X, et al. Hydrophilic, red-emitting, and thermally activated delayed fluorescence emitter for time-resolved luminescence imaging by mitochondrion-induced aggregation in living cells.* Adv Sci. *2019;6:1801729. Copyright @ 2019, Wiley-VCH.*

non-covalent interactions between the TPP cation and tetraphenylborate anion cause NID-TPP aggregation which induces the strong emission.[60] The TRLI process was driven in HeLa cells and the red TADF emission of probe **(29)** was explicitly investigated in mitochondria with low background signals by imaging inside the specified time domain (Fig. 1A–D). This study discovered that TADF-based hydrophilic luminophores and organelle-induced aggregation are useful for time resolved luminescence imaging and two-photon luminescence imaging in oxygenated environments and organisms.[60]

5. Future prospectus

In this chapter, we have summarized the synthesis and applications of AIEgens in mitochondria targeting. Most of the summarized probes are based on TPE, TPA, TPA and some miscellaneous compounds. Mitochondria is one of the cellular organelles to eukaryotic cells, which have an important role in energy production, ROS generation and cellular signaling. The high photostability and biocompatibility make AIEgenes a great candidate for tracking the morphology of mitochondria. These AIEgen probes are used for cancer cell detections, imaging, chemotherapy and photodynamic therapy. These probes have future potential for biological applications.

Acknowledgments

S.V.B. (IICT) is thankful to The Director CSIR-IICT for providing necessary facilities and financial support under the project BRNS No. 58/14/01/2020-BRNS/37047. IICT Commun. No. IICT/Pubs./2021/111. M.R.B. thanks UGC, New Delhi for senior research fellowship.

References

1. Luo J, Xie Z, Xie Z, et al. Aggregation-induced emission of 1-methyl-1,2,3,4,5-pentaphenylsilole. *Chem Commun.* 2001;18:1740–1741.
2. Hu F, Liu B. Organelle-specific bioprobes based on fluorogens with aggregation-induced emission (AIE) characteristics. *Org Biomol Chem.* 2016;14:9931–9944.
3. Duchen MR. Mitochondria and calcium: from cell signalling to cell death. *J Physiol.* 2000;529:57–68.
4. Susin SA, Lorenzo HK, Zamzami N, et al. Molecular characterization of mitochondrial apoptosis-inducing factor. *Nature.* 1999;397:441–446.
5. Osellame LD, Blacker TS, Duchen MR. Cellular and molecular mechanisms of mitochondrial function. *Best Pract Res Clin Endocrinol Metab.* 2012;26:711–723.
6. Silva JP, Kohler M, Graff C, et al. Impaired insulin secretion and β-cell loss in tissue-specific knockout mice with mitochondrial diabetes. *Nat Genet.* 2000;26:336–340.
7. Andersson SGE, Zomorodipour A, Andersson JO, et al. The genome sequence of Rickettsia prowazekii and the origin of mitochondria. *Nature.* 1998;396:133–140.
8. Anderson S, Bankier AT, Barrell BG, et al. Sequence and organization of the human mitochondrial genome. *Nature.* 1981;290:457–465.
9. Wang C, Youle RJ. The role of mitochondria in apoptosis. *Annu Rev Genet.* 2009;43:95–118.
10. Sreedhar A, Aguilera-Aguirre L, Singh KK. Mitochondria in skin health, aging, and disease. *Cell Death Dis.* 2020;11:444.
11. Park D, Lee S, Min KT. Techniques for investigating mitochondrial gene expression. *BMB Rep.* 2020;53:3–9.
12. Huang Y, Zhang G, Zhao R, Zhang D. Aggregation-induced emission luminogens for mitochondria-targeted cancer therapy. *ChemMedChem.* 2020;15:2220–2227.
13. Li J, Wang J, Li H, Song N, Wang D, Tang BZ. Supramolecular materials based on AIE luminogens (AIEgens): construction and applications. *Chem Soc Rev.* 2020;49:1144–1172.
14. Zhang CJ, Hu Q, Feng G, et al. Image-guided combination chemotherapy and photodynamic therapy using a mitochondria-targeted molecular probe with aggregation-induced emission characteristics. *Chem Sci.* 2015;6:4580–4586.
15. Kang M, Zhang Z, Song N, et al. Aggregation enhanced theranostics: AIE sparkles in biomedical field. *Aggregate.* 2020;1:80–106.
16. Mei J, Leung NLC, Kwok RTK, Lam JWY, Tang BZ. Aggregation-induced emission: together we shine, united we soar! *Chem Rev.* 2015;115:11718–11940.
17. Gao M, Sim CK, Leung CWT, et al. A fluorescent light-up probe with AIE characteristics for specific mitochondrial imaging to identify differentiating brown adipose cells. *Chem Commun.* 2014;50:8312–8315.

18. Zhang W, Kwok RTK, Chen Y, et al. Real-time monitoring of the mitophagy process by a photostable fluorescent mitochondrion-specific bioprobe with AIE characteristics. *Chem Commun.* 2015;51:9022–9025.
19. Zhao Z, Chan CYK, Chen S, et al. Using tetraphenylethene and carbazole to create efficient luminophores with aggregation-induced emission, high thermal stability, and good hole-transporting property. *J Mater Chem.* 2012;22:4527–4534.
20. Gu X, Zhao E, Zhao T, et al. Mitochondrion-specific photoactivatable fluorescence turn-on AIE-based bioprobe for localization super-resolution microscope. *Adv Mater.* 2016;28:5064–5071.
21. Rust MJ, Bates M, Zhuang X. Sub-diffraction-limit imaging by stochastic optical reconstruction microscopy (STORM). *Nat Methods.* 2006;3:793–795.
22. Qian J, Tang BZ. AIE luminogens for bioimaging and theranostics: from organelles to animals. *Chem.* 2017;3:56–91.
23. Zhao N, Li M, Yan Y, et al. Tetraphenylethene-substituted pyridinium salt with multiple functionalities: synthesis, stimuli-responsive emission, optical waveguide and specific mitochondrion imaging. *J Mater Chem C.* 2013;1:4640–4646.
24. Yang H, Zhuang J, Li N, et al. Efficient near-infrared photosensitizer with aggregation-induced emission characteristics for mitochondria-targeted and image-guided photodynamic cancer therapy. *Mater Chem Front.* 2020;4:2064–2071.
25. Leung CWT, Hong Y, Chen S, Zhao E, Lam JWY, Tang BZ. A photostable AIE luminogen for specific mitochondrial imaging and tracking. *J Am Chem Soc.* 2013;135:62–65.
26. Zhao X, Chen Y, Niu G, et al. Photostable PH-sensitive near-infrared aggregation-induced emission luminogen for long-term mitochondrial tracking. *ACS Appl Mater Interfaces.* 2019;11:13134–13139.
27. Zhao N, Chen S, Hong Y, Tang BZ. A red emitting mitochondria-targeted AIE probe as an indicator for membrane potential and mouse sperm activity. *Chem Commun.* 2015;51:13599–13602.
28. Li Y, Liu T, Liu H, Tian MZ, Li Y. Self-assembly of intramolecular charge-transfer compounds into functional molecular systems. *Acc Chem Res.* 2014;47:1186–1198.
29. Zhao N, Yang Z, Lam JWY, et al. Benzothiazolium-functionalized tetraphenylethene: an AIE luminogen with tunable solid-state emission. *Chem Commun.* 2012;48:8637–8639.
30. Qin W, Ding D, Liu J, et al. Biocompatible nanoparticles with aggregation-induced emission characteristics as far-red/near-infrared fluorescent bioprobes for in vitro and in vivo imaging applications. *Adv Funct Mater.* 2012;22:771–779.
31. Lim SY, Hong KH, Kim DI, Kwon H, Kim HJ. Tunable heptamethine-azo dye conjugate as an NIR fluorescent probe for the selective detection of mitochondrial glutathione over cysteine and homocysteine. *J Am Chem Soc.* 2014;136:7018–7025.
32. Yang X, Wang Q, Hu P, et al. Achieving remarkable and reversible mechanochromism from a bright ionic AIEgen with high specificity for mitochondrial imaging and secondary aggregation emission enhancement for long-term tracking of tumors. *Mater Chem Front.* 2020;4:941–949.
33. Liang Y, Li Y, Wang H, Dai H. Strongly coupled inorganic/nanocarbon hybrid materials for advanced electrocatalysis. *J Am Chem Soc.* 2013;135:2013–2036.
34. Hu Q, Gao M, Feng G, Liu B. Mitochondria-targeted cancer therapy using a light-up probe with aggregation-induced-emission characteristics. *Angew Chem Int Ed.* 2014;53:14225–14229.
35. Wang Z, Gu Y, Liu J, et al. Novel pyridinium modified tetraphenylethene: AIE-activity, mechanochromism, DNA detection and mitochondrial imaging. *J Mater Chem B.* 2018;6:1279–1285.

36. Zhao E, Deng H, Chen S, et al. Dual functional AEE fluorogen as a mitochondrial-specific bioprobe and an effective photosensitizer for photodynamic therapy. *Chem Commun*. 2014;50:14451–14454.
37. Yuan Y, Kwok RTK, Feng G, et al. Rational design of fluorescent light-up probes based on an aie luminogen for targeted intracellular thiol imaging. *Chem Commun*. 2014; 50:295–297.
38. Mei J, Huang Y, Tian H. Progress and trends in AIE-based bioprobes: a brief overview. *ACS Appl Mater Interfaces*. 2018;10:12217–12261.
39. Situ B, Ye X, Zhao Q, et al. Identification and single-cell analysis of viable circulating tumor cells by a mitochondrion-specific AIE bioprobe. *Adv Sci*. 2020;7:1902760.
40. Guo B, Wu M, Shi Q, et al. All-in-one molecular aggregation-induced emission theranostics: fluorescence image guided and mitochondria targeted chemo-and photo-dynamic cancer cell ablation. *Chem Mater*. 2020;32:4681–4691.
41. Liu J, Liang J, Wu C, Zhao Y. A doubly-quenched fluorescent probe for low-background detection of mitochondrial H_2O_2. *Anal Chem*. 2019;91:6902–6909.
42. Saha PC, Chatterjee T, Pattanayak R, et al. Targeting and imaging of mitochondria using near-infrared cyanine dye and its application to multicolor imaging. *ACS Omega*. 2019;4:14579–145888.
43. Zhang T, Li Y, Zheng Z, et al. In situ monitoring apoptosis process by a self-reporting photosensitizer. *J Am Chem Soc*. 2019;141:5612–5616.
44. Liu Y, Li K, Wu MY, Liu YH, Xie YM, Yu XQ. A mitochondria-targeted colorimetric and ratiometric fluorescent probe for biological SO_2 derivatives in living cells. *Chem Commun*. 2015;51:10236–10239.
45. Wang T, Shah I, Yang Z, et al. Incorporating thiourea into fluorescent probes: a reliable strategy for mitochondrion-targeted imaging and superoxide anion tracking in living cells. *Anal Chem*. 2020;92:2824–2829.
46. Zhuang W, Yang L, Ma B, et al. Multifunctional two-photon aie luminogens for highly mitochondria-specific bioimaging and efficient photodynamic therapy. *ACS Appl Mater Interfaces*. 2019;11:20715–20724.
47. Wang H, Huff TB, Zweifel DA, et al. In vitro and in vivo two-photon luminescence imaging of single gold nanorods. *Proc Natl Acad Sci U S A*. 2005;102:15752–15756.
48. Luo J, Xie Z, Xie Z, et al. Aggregation-induced emission of 1-methyl-1,2,3,4,5-pentaphenylsilole. *Chem Commun*. 2001;18:1740–1741.
49. Chen J, Law CCW, Lam JWY, et al. Synthesis, light emission, nanoaggregation, and restricted intramolecular rotation of 1,1-substituted 2,3,4,5-tetraphenylsiloles. *Chem Mater*. 2003;15:1535–1546.
50. Zhou J, Liu Z, Li F. Upconversion nanophosphors for small-animal imaging. *Chem Soc Rev*. 2012;41:1323–1349.
51. (a) Zheng Y, Ding Y, Ren J, Xiang Y, Shua Z, Tong A. Simultaneously and selectively imaging a cytoplasm membrane and mitochondria using a dual-colored aggregation-induced emission probe. *Anal Chem*. 2020;92:14494–14500. (b) Wang X, Fan L, Wang S, et al. Real-time monitoring mitochondrial viscosity during mitophagy using a mitochondria-immobilized near-infrared aggregation-induced emission probe. *Anal Chem*. 2021;6:3241–3249.
52. Ding WX, Yin XM. Mitophagy: mechanisms, pathophysiological roles, and analysis. *Biol Chem*. 2012;393:547–564.
53. Yu CYY, Zhang W, Kwok RTK, Leung CWT, Lam JWY, Tang BZ. A photostable AIEgen for nucleolus and mitochondria imaging with organelle-specific emission. *J Mater Chem B*. 2016;4:2614–2619.
54. Chen J, Law CCW, Lam JWY, et al. Synthesis, light emission, nanoaggregation, and restricted intramolecular rotation of 1,1-substituted 2,3,4,5-tetraphenylsiloles. *Chem Mater*. 2003;15:1535–1546.

55. Chen Y, Ai W, Guo X, et al. Mitochondria-targeted polydopamine nanocomposite with AIE photosensitizer for image-guided photodynamic and photothermal tumor ablation. *Small.* 2019;15:1–14.
56. Jiang Y, Pu K. Multimodal biophotonics of semiconducting polymer nanoparticles. *Acc Chem Res.* 2018;51:1840–1849.
57. Zhang J, Wang Q, Guo Z, et al. High-fidelity trapping of spatial–temporal mitochondria with rational design of aggregation-induced emission probes. *Adv Funct Mater.* 2019; 29:1–11.
58. Hu Q, Gao M, Feng G, Liu B. Mitochondria-targeted cancer therapy using a light-up probe with aggregation-induced-emission characteristics. *Angew Chem Int Ed.* 2014;53: 14225–14229.
59. Murphy MP, Smith RAJ. Targeting antioxidants to mitochondria by conjugation to lipophilic cations. *Annu Rev Pharmacol Toxicol.* 2007;47:629–656.
60. Ni F, Zhu Z, Tong X, et al. Hydrophilic, red-emitting, and thermally activated delayed fluorescence emitter for time-resolved luminescence imaging by mitochondrion-induced aggregation in living cells. *Adv Sci.* 2019;6:1801729.

CHAPTER EIGHT

AIE materials for nucleus imaging

Ankit Singh, Dhara Chaudhary, Aishwarya P. Waghchoure, Ravi N. Kalariya, and Rajesh S. Bhosale*

Department of Chemistry, School of Science, Indrashil University, Mehsana, India
*Corresponding author: e-mail address: rajeshbhosale24@gmail.com

Contents

1. Introduction	205
2. AIEgens for nucleus imaging	208
2.1 Nucleus imaging	208
2.2 Cancer cell detection and its nucleolus imaging using AIE probes	211
2.3 AIE probes for nucleic acid sensing and imaging	213
3. Conclusion and future prospects	216
Acknowledgments	216
References	216

Abstract

Emergence of a captivating phenomenon aggregation induced emission (AIE) in the early years of 21st century attracted worldwide researchers. In the last two decades various novel AIE active biocompatible small molecules, macromolecules and polymers have been developed for diverse biomedical applications. Imaging of specific organelle such as mitochondria, ribosomes, nuclei and many others play important in the controlling and successful treatment of various diseases. Conventional luminescent probe molecules used in the imaging at cellular or subcellular level exhibit very weak emission on dispersion or on aggregation in aqueous media. AIE luminogens development is indispensable to overcome the notorious aggregation-caused quenching (ACQ) issue inherited by conventional fluorophores. In the present chapter we mostly highlighted over one decade development of various AIE active luminogens utilized for imaging of cell nucleus, nucleon and nucleic acids. The development of those AIE luminogens exhibits promising results in the early diagnosis of cancer diseases.

1. Introduction

Rapid and early diagnosis is crucial in control and successful treatment of many diseases like cancer. Additionally, the bio-distribution, action and kinetics of the therapeutic molecules also play a vital role. The diagnosis, monitoring, and pharmacokinetics rely upon the imaging techniques.

Also, mechanisms of many biological processes can only be elucidated by imaging at cellular or subcellular level. In particular, nucleus imaging becomes more important as a wide variety of cell behaviors are controlled and regulated by the nucleus such as deoxyribonucleic acid (DNA) replication, gene expression etc.[1] In addition to these, nuclear imaging also plays a vital role in diagnosis of many diseases including cancer as the nuclear structure and framework varies remarkably from one cell type to the other.[2,3] Also, it aids better manifestation of behavior of cancer cells which ultimately provides an understanding and basis for development of more accurate diagnostic and therapeutic tools for cancer.[4]

Radioactive probes are the most primary tools used in bio-imaging, sensing and therapy[5–8] but, they do come with various demerits. Firstly, the radioactive reagents are expensive and the setup requires high-cost maintenance[9,10] which restricts their use in rapid and high-output studies.[11] Moreover, the radiopharmaceuticals are toxic, less bio-compatible and can induce severe mutations as Gamma radiations are harmful. On the other hand, fluorescence dyes have gained significant interest in the last few years for imaging and sensing purposes thanks to its high sensitivity and resolution.[12] For nucleus imaging, use of hoechst dyes, ethidium bromide, acridinium salts, cyanin compounds, ruthenium complexes, propidium iodide, etc. as fluorescence dyes have been reported.[13–20] But, at the same time, these fluorophores are normally carcinogenic with probable health hazards and they also lack tissue penetration, photostability, small Stokes shift, low fluorescence quantum yield, low water solubility which, checks their applications in long term imaging.[21–23] Besides radioisotopes and fluorophores, Quantum Dots (QDs) have been widely explored for imaging purposes but they possess a handful of challenges such as inherent cytotoxicity, low aqueous solubility, non- specific bio-distribution, etc.[24–27]

On the top of these, the fluorophores, for example, naphthalene and QDs also suffer through aggregated caused quenching (ACQ) effect which apparently limits their application. The fluorophores and the QDs generally show strong fluorescence at lower concentrations and in solution state but in aggregated state or in the solid state there is a significant decrease in the fluorescence intensity or even complete quenching. This phenomenon is termed as aggregated caused quenching effect. In the aggregated state, ACQ effect mostly emerge from short-range molecular interactions, such as π–π stacking, which originates due to the planar structure of the molecules (Fig. 1), leading to non-radiative relaxation pathways which ultimately reduces or quenches the fluorescence.[28]

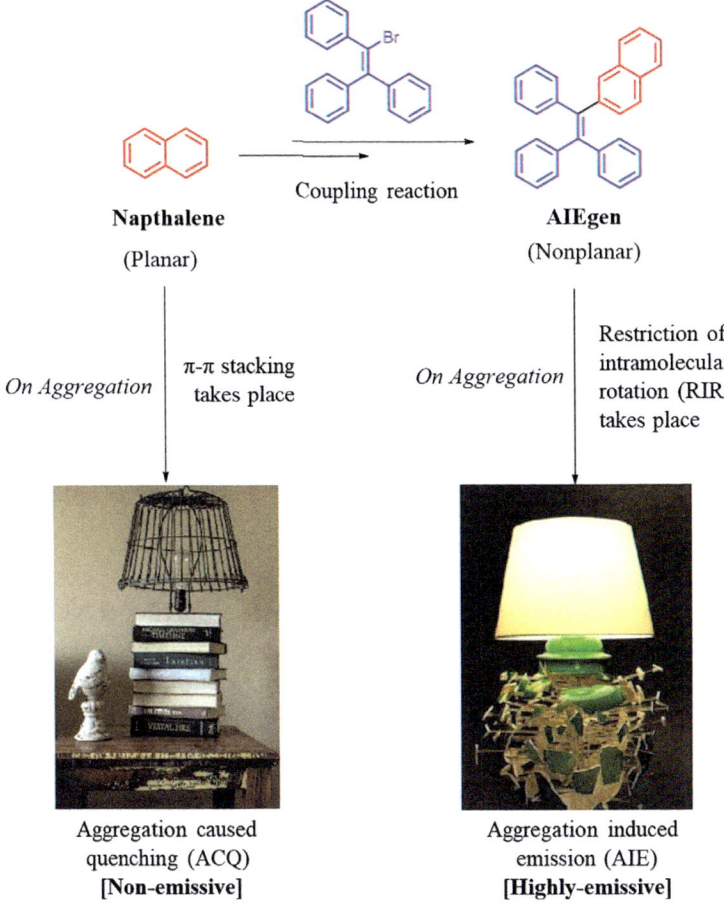

Fig. 1 Schematic diagram of ACQ and AIE effect.

Hence, there is a high demand to develop novel fluorescent probes with low cytotoxicity, good bio-compatibility, solubility, and to overcome the ACQ effect. In 2001, Tang et al. reported a captivating phenomenon aggregation induced emission (AIE).[29] The AIE active fluorophores known as AIEgens are well distinguished from the conventional ACQ active fluorophores, they show high fluorescence intensity upon aggregation and in solid state which makes them ideal candidates for long term nucleus imaging. This is due to the fact that the AIE active fluorophores when aggregated undergoes the phenomena of restriction of intramolecular rotation (RIR) which in turn activates the radiative decay pathways and become highly emissive (Fig. 1).[30] In this chapter, we have summarized the development of AIE active fluorophores and its employment in nucleus imaging in the last two decades.

2. AIEgens for nucleus imaging

Last two decades evolution of AIE concept attracted several worldwide researchers to develop novel AIE active materials for various biomedical applications. Design and development of specific AIEgens for the nucleus imaging is challenging task. Till date small numbers of literatures are available on AIEgens used for cellular nucleus imaging in the normal and cancer cells. In the following sections we systematically summarized those important literatures.

2.1 Nucleus imaging

Yong Cheng et al. in 2013, reported a dual targeted novel multifunctional peptide conjugated fluorescent probe with AIE properties **(cNGR-CPP-NLS-RGD-PyTPE, TCNTP)** for long term nucleus imaging in the cancer cells. **TCNTP** mainly comprised of a multifunctional peptide delivery system cobbled with **Py-TPE** moiety, an AIE fluorescence agent (Fig. 2A).[31] Talking about the peptide delivery system, it consisted of four components for specific targeting of the nucleus of cancer cells. Firstly, **RGD**, a targeting peptide and **CNGR**, another targeting peptide motif but with cyclic structure were used for specific binding of the **TCNTP** to the cancer cell surface. Once, the **TCNTP** was bound to the cell surface, its emission was switched on. Both the peptides, **RGD** and **CNGR** had high affinity toward integrin $\alpha_v\beta_3$ and **CD13** respectively, which are highly expressed in cancer cells. Additionally, a guanidium rich cell-penetrating peptide **(CPP)**, which augmented the solubility of **TCNTP** in cellular environment and improved its ability to pass through cell membrane. Besides, **RGD**, **CDB** and **CPP**, a nuclear localization signal **(NLS)** was introduced for enhanced transportation of the probe into the cell nucleus. The mechanism of delivery of **TCNTP** into the cell nucleus is shown in the Fig. 2B.

Staining the A375 cancer cells with **TCNTP** revealed that it could specifically light up the cell nucleus, thanks to targeting delivery peptide. Also, the fluorescent stains of **TCNTP** transmitted to the daughter cells and were visible even after the 10th culture cycle that too with low cyto-toxicity.

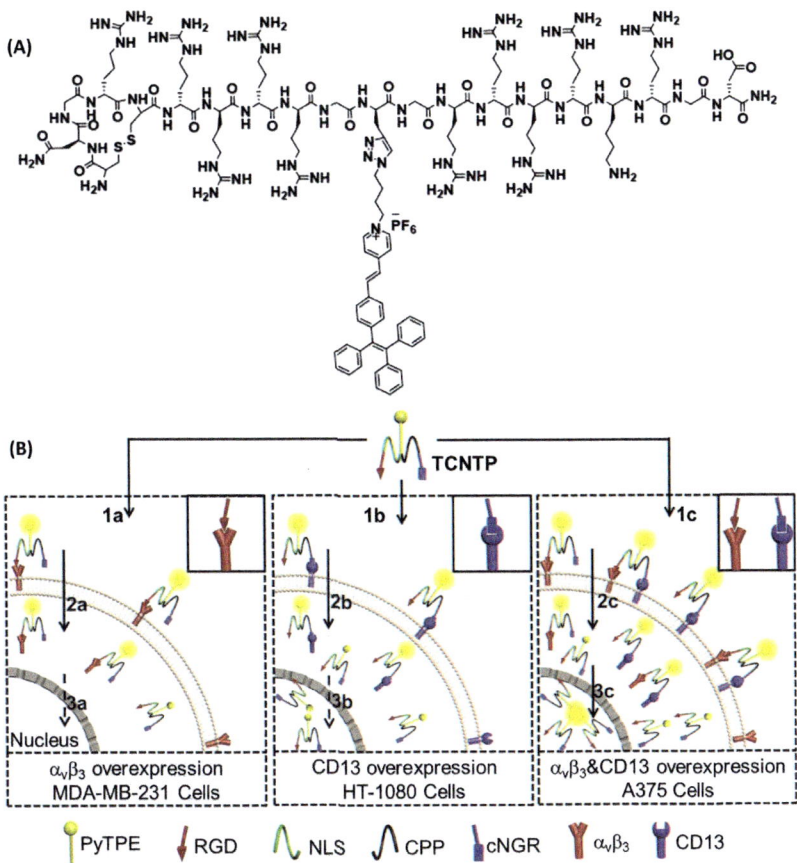

Fig. 2 (A) Chemical structure of AIE active TCNTP probe; (B) Schematic presentation of TCNTP as an emissive nuclear penetrating molecule in integrin $\alpha_v\beta_3$ and CD13 over-expression living cells. The integrin $\alpha_v\beta_3$ and CD13 overexpression cells would bind more TCNTP in cytomembrane which could enter into nucleus easily with further incubation.

Another AIE dots probe **TPA-AN-TPM@PS-PVP** emitting red fluorescence was employed by Tian Wenjing and coworkers for cytoplasm and nucleus imaging.[32] The probe mainly consisted of a core (2-(2,6-bis((E)-2-(5-(N,N-bis(4-((E)-2-(10-((E)-4-(diphenylamino) styryl)anthracen-9-yl)vinyl)phenyl)aniline-4-yl)thiophen-2-yl)vinyl)-4H-pyran-4-ylidene) malononitrile) **TPA-AN-TPM** (Fig. 3A), an AIE molecule encapsulated in a bio-compatible poly(styrene)-poly(4-vinylpyridine) polymer micelle **PS-PVP** which showed a quantum yield of 12.9%. The probe was subjected for the imaging of HELA cells and was studied using CLSM which

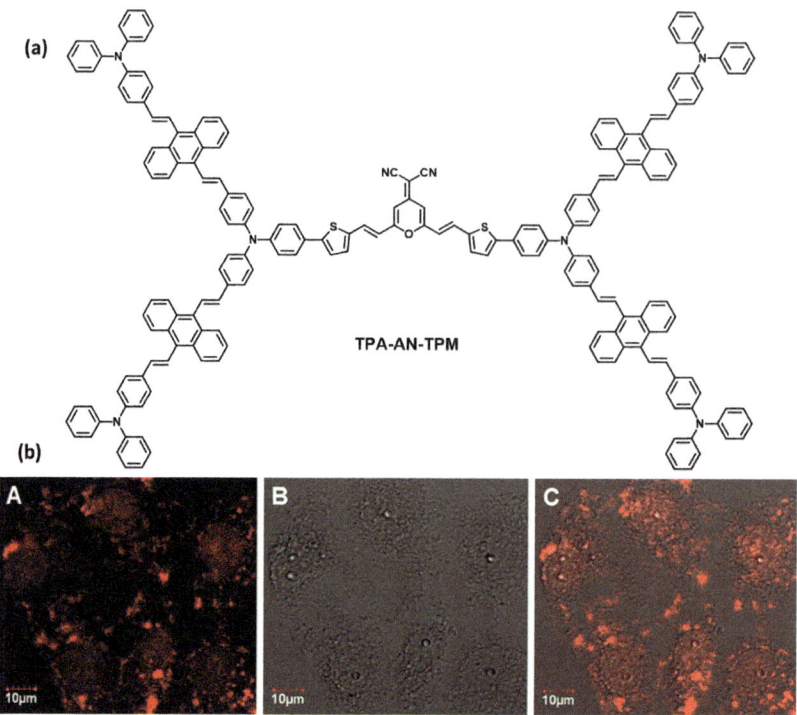

Fig. 3 (A) Chemical structure of AIE active TPA-AN-TPN probe, CLSM images of HeLa cells after incubation with TPA-AN-TPM@PS-PVP (with a fluorogen loading of 10%) for 16 h at 37 °C. (A) Fluorescence image; (B) bright field image; (C) overlay of (A) and (B). Concentration of AIE dots: 0.15 mg mL^{-1}.

revealed that it stained both the cytoplasm and nucleus but more intense fluorescence was observed in the nucleus (Fig. 3B). Also, the AIE dots were bio-compatible with low cyto-toxicity and good aqueous stability.

In a new kind of study, Dubin Chao et al. fabricated a cost effective carbazole modified tetrapyridine Zn (II) complex (**CZtpyZn**) with AIE characteristics and successfully utilized the same for the specific detection of PPi in aqueous solutions of pH 6–8. The Zn complex was weakly emissive but once it was assembled with PPi, it produced a nano supramolecular assembly which, further formed nano aggregates to yield strong fluorescence (Fig. 4A) and the fluorescence intensity increased linearly with the concentration of PPi.[33] In the view of the above strategy and taking into the account of the fact that PPi is universally present in all biological fluids, the same complex was employed for the nucleus staining in HELA cells. Results showed that the **CZtpyZn** complex primarily stained the nucleus as there was very less fluorescence from the cytoplasm. The cells staining

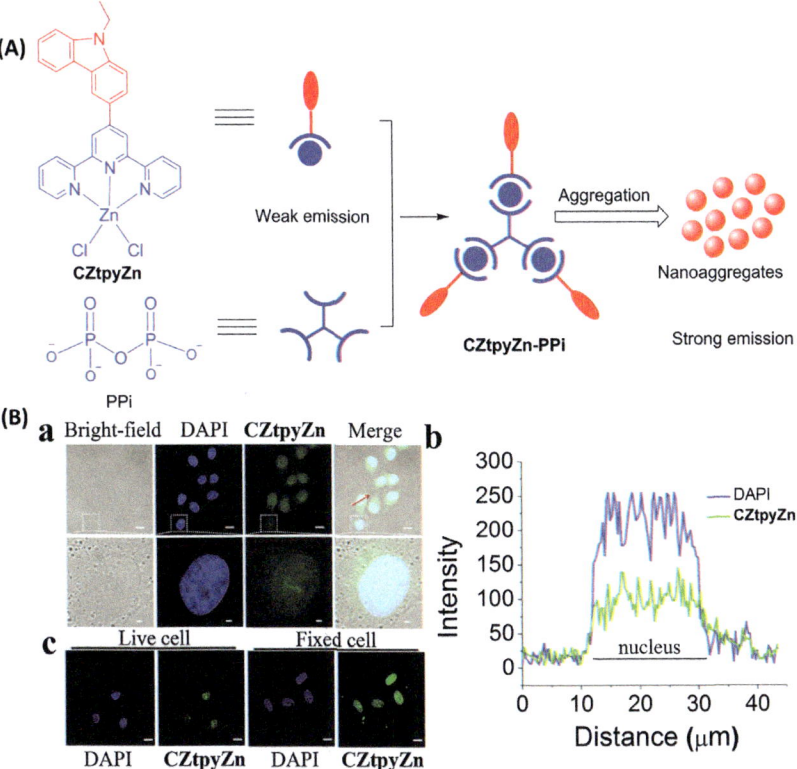

Fig. 4 (A) Supramolecular assembly (CZtpyZn–PPi) of terpyridine–Zn(II) complex (CZtpyZn) coordinates with PPi which on self-assembly form aggregates with strong emission; (B) Nucleus staining with CZtpyZn (A) The confocal fluorescence images of HeLa cells incubated with CZtpyZn (5 μM) for 30 min and then further incubated with DAPI (5 μg mL^{-1}) for 10 min. Blue channel for DAPI excited at 405 nm and green channel for CZtpyZn excited at 488 nm. (B) Fluorescence profile of intensity across the red line in (A). (C) The confocal fluorescence images of living HeLa cells and fixed HeLa cells with CZtpyZn. Scale bar: 20 μm.

with 4,6-diamidino-2-phenylindole (DAPI) confirmed that **CZtpyZn** stained the nucleus more effectively than the cytoplasm but the fluorescence intensity was weaker than DAPI (Fig. 4B). Further, the MTT assays revealed that the complex possessed very low cyto-toxicity and more bio-compatibility, making this low-cost probe a strong candidate for nucleus imaging.

2.2 Cancer cell detection and its nucleolus imaging using AIE probes

In addition to the nucleus imaging of cells using AIE probes, few studies were reported which made use of the AIE luminogens for the detection of cancer cells as well as the imaging of its nucleolus. A novel AIE based nanoprobe was designed by Tang Gao and group which was able to differentiate normal healthy cells and cancerous cells. Also, it specifically targeted and the stained the nucleolus of the cancer cells for a long period of time.[34] They synthesized a small bola type water soluble molecule 4-(1-(4-chlorophenyl)-4,5-diphenyl-1H-imidazol-2-yl)-1-(12-(pyridin-1-ium-1-yl)dodecyl)pyridin-1-ium bromide denoted as **DPMPB**, it is capable of self-assembling into positively charged nanoparticles (**DPMPB-FONs**) which emitted very weak fluorescence in aqueous solution. The positive charge on the pyridinium ion enhanced its ability to penetrate into the tumor cells via charge mediated *endo*-cytosis (Fig. 5A). Also, the **DPMPB** had

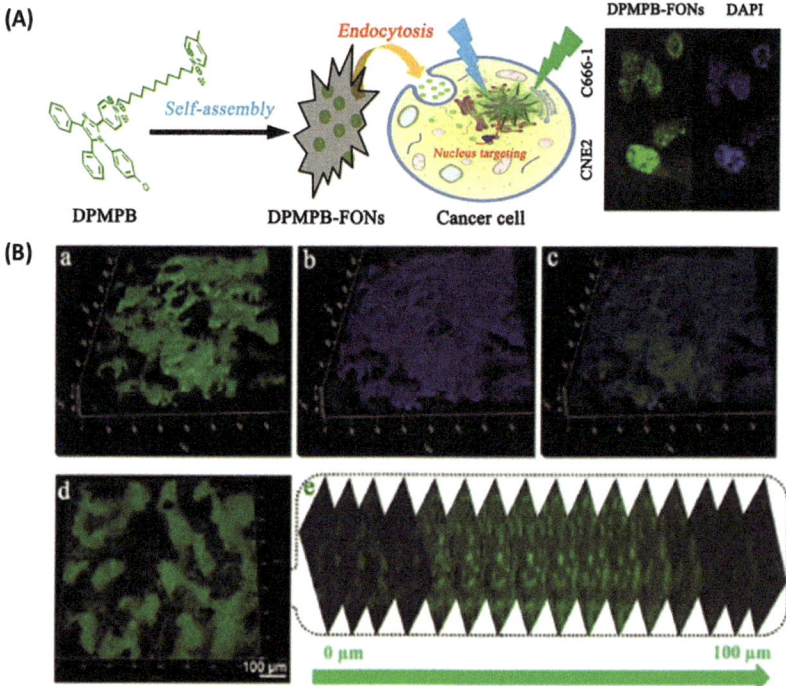

Fig. 5 (A) Chemical structure of fluorescent nano probe DPMPB used for selective cancer cells nucleus imaging; (B) Schematic presentation (A) 3D reconstruction fluorescence images of breast cancer tissue stained with 5 μM of DPMPB-FONs in two-photon mode by collecting the emissions at 480–530 nm upon excitation at 760 nm. (B) 3D projection fluorescence images of breast cancer tissue stained with 1 μM DAPI by collecting the emissions at 450–470 nm upon excitation at 358 nm. (C) merged image of (A) and (B). (D) 3D reconstruction from 50 confocal Z-scan two-Photon excited imaging sections at depth of 0–100 μM (E) Sectional TPM images of breast cancer tissue labeled with DPMPB-FONs (5 μM) from the mucosal surface (0 μm) to 100 μm depth. Scale bar: 100 μm.

excellent optical properties and low cyto-toxicity. Above all, using two-photon imaging, **DPMPB-FONs** was efficiently distinguishing the breast cancer cells and para-carcinoma tissues clinical samples with very high tissue penetration depth (Fig. 5B) which indicates the potential of this AIE probe as a guiding-agent in surgery.

In a similar study, an AIE based luminogens, responsive to the nuclear-density of the cell was developed to distinguish tumor cells from normal cells. Bo Situ et al. prepared a small positively charged molecular probe **MASPB** with AIE features.[35] Like DPMPB, **MASPB** too was mainly localized in and brightly lit up the nucleolus of the cancer cells. However, the mechanism of **MASPB** here was a bit novel. Various experiments such as imaging of WBCs, HELA cells treated with RNAse and DNAse, and different phases of cell cycle of lymphocytes with **MASPB** revealed that the structural density of the nucleus was responsible for the fluorescence of **MASPB** (Fig. 6), making it a Nuclear-Density responsive AIE bio-probe. Moreover, **MASPB** was used to stain malignant pleural effusion (MPE) samples from lung cancer patients to detect rare tumor cells. Surprisingly, **MASPB** was able to recognize and specifically stain the malignant cells in the MPE samples which made the tumor cells to be easily distinguished even when they were present in very small proportion in high white blood cells (WBC) background.

2.3 AIE probes for nucleic acid sensing and imaging

Besides nucleus and nucleolus of the cell, nucleic acids also are of great biological importance. Generally, Alexa-647-azide dye is used along with

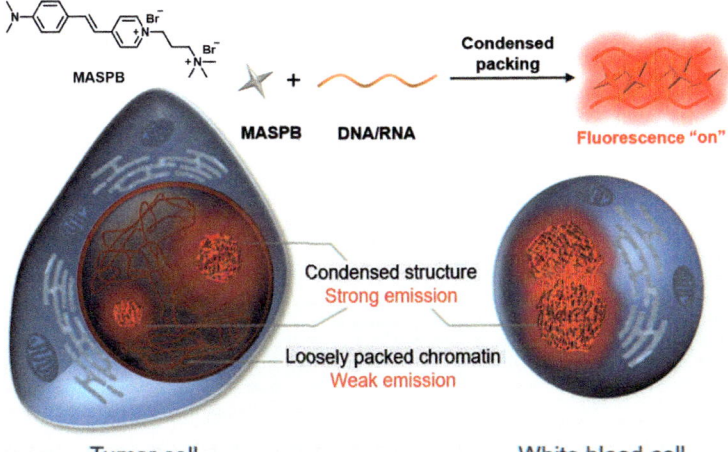

Fig. 6 Schematic presentation of chemical structure of MASPB and the sensing mechanism.

5-ethynyl-2′-deoxyuridine (EdU) for the detection of DNA synthesis but it as suffers from ACQ effect, development of new AIE based fluorophores has gained a lot of interest. In this section of the chapter we have synopsized AIE probes developments for nucleic acids sensing and imaging. Yueyue Zhao and group in 2014 reported two azide functionalized tetraphenylethene (TPE)-derivatives with AIE characteristics for precise and highly responsive detection of DNA synthesis during S-phase and cell proliferation.[36] The detection mechanism two AIEgens **TPE-Py-N$_3$** and **Cy-Py-N$_3$** was based on the EdU- assay (Fig. 7). The DNA synthesis was studied using **TPE-Py-N$_3$** in proliferating HELA cells. As soon as the DNA synthesis was initiated, the cells took up EdU from the solution to which the **TPE-Py-N$_3$** was further bound making the DNA as well as the nucleus fluorescent and visible under a fluorescence microscopy. Also, the fluorescence intensity increased with the concentration of EdU and **TPE-Py-N$_3$** and two significant peaks were observed at G_1 phase and G_{2M} phase which confirmed the S-phase of the cell cycle. Additionally, the AIEgen showed better photo-stability, high photo-bleaching resistance as compared to conventional Alexa-azide dye.

Xiaowei Xu et al. in a similar study, fabricated an AIE fluorescence probe **FcPy**, which is a derivative of *p*-phenylenediacetonitrile for detection and

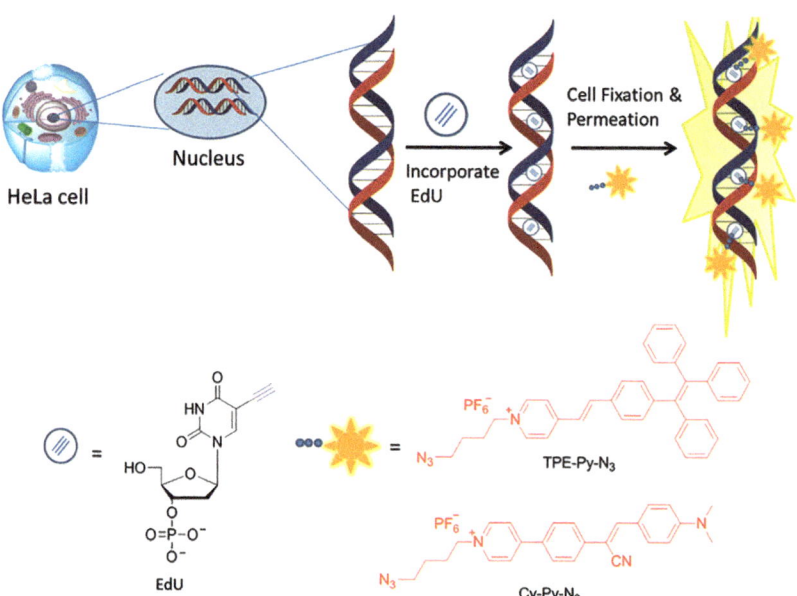

Fig. 7 Chemical structure of EdU, TPE-Py-N$_3$ and Cy-Py-N3; EdU assay based DNA synthesis detection by AIEgens.

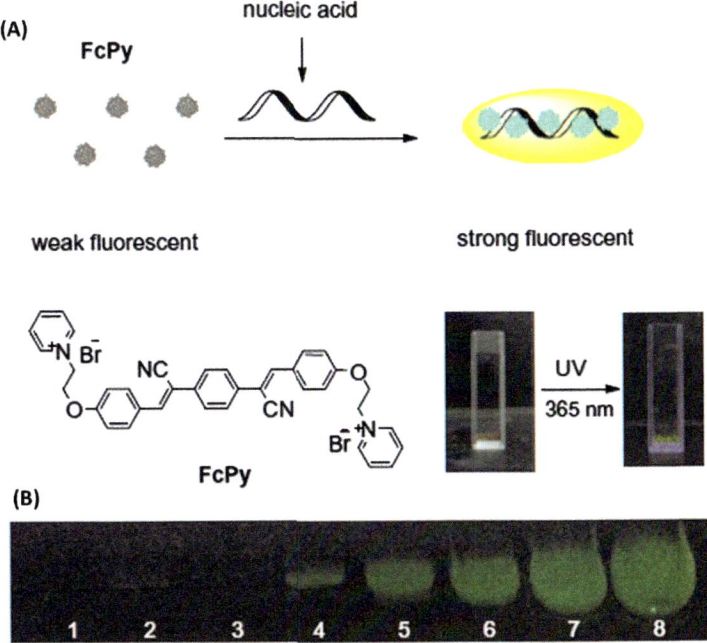

Fig. 8 (A) Chemical structure and schematic presentation of the FcPy system used for nucleic acid detection via fluorescence "Turn-On" mechanism; (B) Single-strand DNA staining in PAGE by FcPy (100 μM). Concentrations of DNA (75 nt) in lanes 1–8: 0, 0.25, 0.5, 1.0, 10.0, 25.0 and 50.0 μg (from left to right). Staining time: 20 min.

quantification of nucleic acids in solutions and cellular systems. **FcPy** being cationic in its aqueous solution, it readily adhered to the negatively charged nucleic acid, thereafter emitting strong fluorescence (Fig. 8A).[37]

Firstly, single stranded DNA was detected, which triggered the fluorescence of **FcPy** and its intensity increased with the length of DNA sequence (Fig. 8B) but, the **FcPy** showed no selectivity between single and double stranded DNA in the essence of the fact that the fluorescence intensity was almost doubled in the case of double stranded DNA. Also, the fluorescence intensity was enhanced at lower concentration and it gradually saturated at higher concentrations, but, for ribonucleic acid (RNA) the fluorescence slightly enhanced at lower concentrations and increased rapidly with increase in the concentration, indicating its ability to detect a wider range of RNA concentrations as compared to DNA. Additionally, **FcPy** was successfully utilized for the nucleus imaging in HELA cells as well.

3. Conclusion and future prospects

Taking into consideration, the importance of nucleus imaging in diagnostics and therapy, and the limitations associated with various imaging probes including the ACQ effect; in this review we have summarized the development of AIE luminogens for nuclear imaging in the last two decades. According to the recent studies and advancements in this field, AIE probes have shown promise across various aspects in nucleus imaging such as good stability, low cyto-toxicity, and high fluorescence in aggregated state etc. More particularly, few AIE probes have been reported that can identify and selectively stain the nucleus of cancer cells, which has triggered more hope and interest among researchers for the development of more advanced cancer-diagnostics tools in future using the AIE enabled fluorescent probes.

Acknowledgments

APW and RNK are acknowledges research SHODH fellowship from Education Department, Gujarat Government. AK, DC and RSB are thankful to Dr. J. S. Yadav Provost & Research Director, Indrashil University for providing research facilities and scientific environment.

References

1. Lanctôt C, Cheutin T, Cremer M, Cavalli G, Cremer T. Dynamic genome architecture in the nuclear space: regulation of gene expression in three dimensions. *Nat Rev Genet.* 2007;8:104–115.
2. Solovei I, Kreysing M, Lanctôt C, et al. Nuclear architecture of rod photoreceptor cells adapts to vision in mammalian evolution. *Cell.* 2009;137:205–207.
3. Woodcock CL, Ghosh RP. Chromatin higher-order structure and dynamics. *Cold Spring Harb Perspect Biol.* 2010;2:a000596.
4. Zink D, Fischer AH, Nickerson JA. Nuclear structure in cancer cells. *Nat Rev Cancer.* 2004;4:677–687.
5. Fernandes RS, Ferreira CA, Soares DCF, et al. Biomed. *Pharmacotherapy.* 2017;95:469–476.
6. Farzin L, Sheibani S, Moassesi ME, Shamsipur M. An overview of nanoscale radionuclides and radiolabeled nanomaterials commonly used for nuclear molecular imaging and therapeutic functions. *J Biomed Mater Res A.* 2019;107:251–285.
7. Pais HL, Alho I, Vendrell I, Mansinho A, Costa L. Radionuclides in oncology clinical practice—review of the literature. *Dalton Trans.* 2017;46:14475–14487.
8. Keinänen O, Brennan JM, Membreno R, et al. Dual radionuclide theranostic pretargeting. *Mol Pharm.* 2019;16:4416–4421.
9. Kim Y-S, Nwe K, Milenic DE, Brechbiel MW, Satz S, Baidoo KE. Synthesis and characterization of $\alpha_v\beta_3$-targeting peptidomimetic chelate conjugates for PET and SPECT imaging. *Bioorg Med Chem Lett.* 2012;22:5517–5522.
10. Dobrucki LW, Sinusas AJ. PET and SPECT in cardiovascular molecular imaging. *Nat Rev Cardiol.* 2010;7:38–47.

11. Maurer HR. Potential pitfalls of [3H] thymidine techniques to measure cell proliferation. *Cell Prolif.* 1981;14:111–120.
12. Wang Y, Xu C, Ow H. Commercial nanoparticles for stem cell labeling and tracking. *Theranostics.* 2013;3:544–560.
13. Feng S, Kim YK, Yang S, Chang Y-T. Discovery of a green DNA probe for live-cell imaging. *Chem Commun.* 2010;46:436–438.
14. Kuruvilla E, Nandajan PC, Schuster GB, Ramaiah D. Acridine-viologen dyads: selective recognition of single-strand DNA through fluorescence enhancement. *Org Lett.* 2008;10:4295–4298.
15. Veale EB, Frimannsson DO, Lawler M, Gunnlaugsson T. 4-Amino-1,8-naphthalimide-based Tröger's bases as high affinity DNA targeting fluorescent supramolecular scaffolds. *Org Lett.* 2009;11:4040–4043.
16. Granzhan A, Ihmels H, Viola G. 9-donor-substituted acridizinium salts: versatile environment-sensitive fluorophores for the detection of biomacromolecules. *J Am Chem Soc.* 2007;129:1254–1267.
17. Deng R, Xie X, Vendrell M, Chang Y-T, Liu X. Intracellular glutathione detection using MnO_2-nanosheet-modified upconversion nanoparticles. *J Am Chem Soc.* 2011; 133:20168–20171.
18. Bucevičius J, Lukinavičius G, Gerasimaitė R. The use of hoechst dyes for DNA staining and beyond. *Chemosensors.* 2018;6(1 – 12):18.
19. Wu D, Shen Y, Chen J, Liu G, Chen H, Yin J. Naphthalimide-modified near-infrared cyanine dye with a large stokes shift and its application in bioimaging. *Chin Chem Lett.* 2017;28:1979–1982.
20. Yagishita F, Tanigawa J-I, Nii C, et al. Fluorescent imidazo[1,5-a]pyridinium salt for a potential cancer therapy agent. *ACS Med Chem Lett.* 2019;10:1110–1114.
21. Li J-B, Liu H-W, Fu T, Wang R, Zhang X-B, Tan W. Recent progress in small-molecule near-IR probes for bioimaging. *Trends Chem.* 2019;1:224–234.
22. Martin RM, Leonhardt H, Cardoso MC. DNA labeling in living cells. *Cytometry A.* 2005;67A:45–52.
23. Resch-Genger U, Grabolle M, Cavaliere-Jaricot S, Nitschke R, Nann T. Quantum dots versus organic dyes as fluorescent labels. *Nat Methods.* 2008;5:763–775.
24. Kumar S. In: Tan M, Wu A, eds. *Nanomaterials for Tumor Targeting Theranostics. A Proactive Clinical Perspective.* 1st ed. World Scientific; 2016.
25. Huang K, Dou Q, Loh XJ. Nanomaterial mediated optogenetics: opportunities and challenges. *RSC Adv.* 2016;6:60896–60906.
26. Pohanka M. Quantum dots in the therapy: current trends and perspectives. *Mini Rev Med Chem.* 2017;17:650–656.
27. Kolmykov O, Coulon J, Lalevée J, Alem H, Medjahdi G, Schneider R. Aqueous synthesis of highly luminescent glutathione-capped Mn_{2+}-doped ZnS quantum dots. *Mater Sci Eng C.* 2014;44:17–23.
28. Bras LL, Chaitou K, Aloïse S, Adamo C, Perrier A. Aggregation-caused quenching versus crystallization induced emission in thiazolo[5,4-b]thieno[3,2-e]pyridine (TTP) derivatives: theoretical insights. *Phys Chem Chem Phys.* 2019;21:46–56.
29. Luo J, Xie Z, Lam JWY, et al. Aggregation-induced emission of 1-methyl-1,2,3,4,5-pentaphenylsilole. *Chem Commun.* 2001;1740–1741.
30. Bhosale RS, Aljabri M, La DD, Bhosale SV, Jones LA, Bhosale SV. Tetraphenylethene derivatives: a promising class of AIE luminogens synthesis, properties, and applications. In: Tang Y, Tang B, eds. *Principles and Applications of Aggregation Induced Emission.* 1st ed. Cham: Springer; 2019. https://doi.org/10.1007/978-3-319-99037-8_9.
31. Cheng Y, Sun C, Ou X, Liu B, Lou X, Xia F. Dual-targeted peptide-conjugated multifunctional fluorescent probe with AIEgen for efficient nucleus-specific imaging and long-term tracing of cancer cells. *Chem Sci.* 2017;8:4571–4578.

32. Wang Z, Yan L, Zhang L, et al. Ultra bright red AIE dots for cytoplasm and nuclear imaging. *Polym Chem*. 2014;5:7013–7020.
33. Chao D, Ni S. Nanomolar pyrophosphate detection and nucleus staining in living cells with simple terpyridine-Zn(II) complexes. *Sci Rep*. 2016;6:26477.
34. Gao T, Wang S, Lv W, et al. A self-assembled nanoprobe for the long term cancer cell nucleus specific staining and two-photon breast cancer imaging. *Chem Commun*. 2018;54:3578–3581.
35. Situ B, He B, Chen X, et al. Fluorescent sensing of nucleus density assists in identifying tumor cells using an AIE luminogens. *Chem Eng J*. 2021;410:128183.
36. Zhao Y, Yu CYY, Kwok RTK, et al. Photostable AIE fluorogens for accurate and sensitive detection of S-phase DNA synthesis and cell proliferation. *J Mater Chem B*. 2015;3:4993–4996.
37. Xu X, Yan S, Zhou Y, et al. A novel aggregation-induced emission fluorescent probe for nucleic acid detection and its applications in cell imaging. *Bioorg Med Chem Lett*. 2014;24:1654–1656.

CHAPTER NINE

Aggregation induced emission molecules for detection of nucleic acids

Rupesh Maurya[a], Gargi Bhattacharjee[a], Nisarg Gohil[a], Khalid J. Alzahrani[b], and Vijai Singh[a,*]

[a]Department of Biosciences, School of Science, Indrashil University, Mehsana, India
[b]Department of Clinical Laboratories Sciences, College of Applied Medical Sciences, Taif University, Taif, Saudi Arabia
*Corresponding author: e-mail addresses: vijaisingh15@gmail.com; vija.singh@indrashiluniversity.edu.in

Contents

1. Introduction	219
2. Mechanism of action of AIE molecules for nucleic acids detection AIE for DNA detection	221
3. AIE for RNA detection	224
4. Conclusion and future perspectives	224
Acknowledgment	224
References	225

Abstract

Aggregation-induced emission (AIE) is an ingenious concept in the field of luminescent molecules. AIE is the energy released in an excited state that in turn is converted into light irrespective of being in either liquid phase or solid phase. Aggregation or crystallization of AIE molecules impedes the free movement of molecules and it resultantly becomes highly fluorescent. It is currently being used for several applications including sensing, diagnostic, protein, DNA or RNA detection, cells and cell organelles imaging. AIEs are highly sensitive and specific for binding with target molecules. In this chapter, we underline different AIE molecules for detection of nucleic acids.

1. Introduction

Fluorescence has brought many advantages to science such as sensitivity and specificity as well as it is also known for its rapidity and ease of operation. It is also a very useful tool for visualization at molecular level and in recording biological mechanisms. Although some inorganic nanoparticles exhibit excellent brightness and photostability, they do show toxicity.

Hence, new kinds of sensors are required which are biocompatible and photostable. During the early years of 21st century, Professor Ben Zhong Tang introduced a very interesting photophysical phenomenon called aggregation induced emission (AIE). In the action of AIE phenomenon, the non-emissive materials are induced to emit light upon its aggregation. AIE was designed and identified as an abnormal phenomenon of certain organicluminophores. Most of the organic molecules have a planar structure. It tends to show higher photoemission activity in solution phase than in solid phase. Organic luminophores molecules possess the ability of free rotations and when these molecules reach an excited state, instead of releasing energy (as light), they are relaxed back down via rotation. However, when these molecules aggregate or crystallize, free rotation gets restricted and they become highly fluorescent and photoemission efficiency significantly increases.[1–4] Hong et al.[1] reported several applications of AIE molecules such as for sensing, use as probes (detection of protein, DNA, RNA), immunoassay markers, reporter for organelles and their locations. It has also been used for fabrication of light emitting diodes.

AIE luminogens (AIEgens) are unique fluorescent molecules which can be utilized for biosensing various molecules such as ions, amino acids, carbohydrates, DNA/RNAs, peptides/proteins, cellular structures and organelles, cancer cells, bacteria and other applications such as killing of pathogens.[5] Currently, AIE is rapidly growing and has gained a lot of scientific interest. It is among the high priority areas for research and development. Number of AIE-based sensors have been developed for detection of metals, chemicals, toxins and more. For example, fluorescence intensity of silver sulfide quantum dots (Ag_2SQDs) was enhanced in presence of rare earth ions using AIE. With the help of those rare ions, fluoride (F^-) was detected in living cells. AIE was also used for bioimaging of F^- in live cells.[6] Metal ions are important for biological processing. These have been selectively and sensitively detected and imaged in living cells using AIE molecules.[7]

Nucleic acids play an incomparable role in life processes such as encoding genetic sequences, storage of genetic information, propagation and metabolism.[8,9] Mutations in genetic material are associated with many diseases.[10,11] Tracing mutation in genetic sequences can significantly help in detecting diseases.[12,13] Synthetic nucleic acids have significant roles in therapeutics,[14] disease management as well as diagnostics.[15] AIEgens functional nucleic acids (AFNAs) probe can detect Hg^{2+} ions. Modification in AFNAs cause inhibition of rotation, which generates fluorescence signals.[9] In this chapter, we present and highlight AIE molecules for detection of nucleic acids to meet several biological, therapeutic and biomedical applications.

2. Mechanism of action of AIE molecules for nucleic acids detection AIE for DNA detection

AIEs have already been used for detection of specific DNA sequences for identification of organisms and disease conditions. As shown in Fig. 1, synthesis of TPE–DNAp for detection of nucleic acid hybridization and DNAp as well as DNAt used as probe DNA that has targeted complementary DNA to form a complex.[16] Xu et al.[17] developed a tetraphenylethene-based sensor for selective detection of Hg^{2+}-DNA complex. In presence of Hg^{2+}, aptamer ssDNA with thymine (T)-rich DNA sequences binds to Hg^{2+} and forms a Hg^{2+}-bridged T-base pair. When, the ssDNA was changed into a hairpin-like structure and was positioned closer, it was observed that the fluorescence intensity and sensitivity for Hg^{2+} detection increased. Hong et al.[18] have developed non-emissive tetraphenylethene derivatives

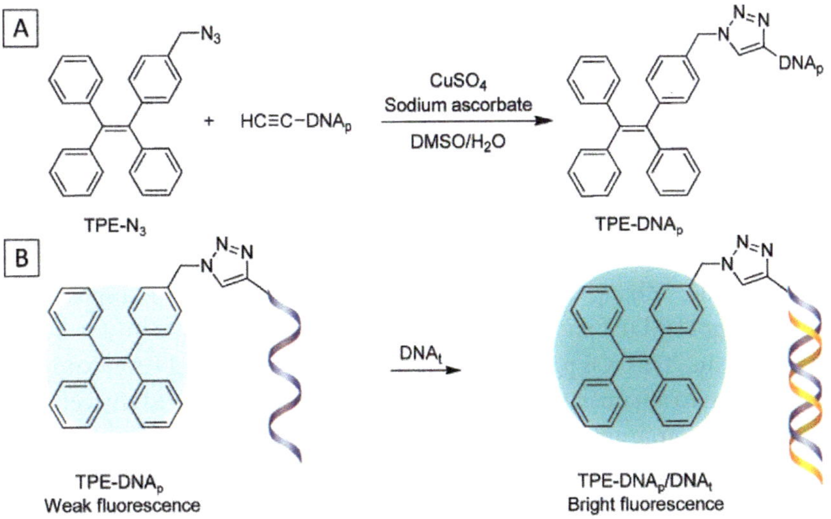

Fig. 1 Synthesis of TPE–DNAp (A) detection of nucleic acid hybridization (B). DNAp and DNAt used as probe DNA and the target complementary DNA to form a complex. *Figure reproduced with permission from Li Y, Kwok RT, Tang BZ, Liu B. Specific nucleic acid detection based on fluorescent light-up probe from fluorogens with aggregation-induced emission characteristics. RSC Adv. 2013; 3: 10135–10138.* © *RSC Advances.*

(TTAPE) which are induced by DNA/RNA for emitting light. It has been used for quantification and visualization of nucleic acids both in liquid and electrophoretic gels. TTAPE has strong specific binding affinity toward DNA that allows to label chromosomes and nuclei in fixed cells. It has shown to aggregate in DNA-rich regions with high photostability effects. A list of AIE for detection of nucleic acid is given in Table 1.

Recently, Hu et al.[31] used AIE with carboxyl—modified tetraphenylethylene (TPE) which is an iconic AIE fluorogen bound with amino-modified DNA with covalent interactions (TPE/DNA). TPE was attached to the outer layer of magnetic beads (MB) through biotin-modified TPE/DNA to form MB/DNA/HA/TPE. These conjugates show better fluorescence response more accurately than conventional conjugated AIE/biomolecules. Under optimal conditions, it was found that fluorescent intensity increased and the detection limit was raised to 1.0×10^{-9} M. Kawamura et al.[32] developed AIE-based facile DNA detection method along with dye-labeled peptide nucleic acid (PNA). It has shown to play role in target recognition and signal emission, and can help to quantify specific target DNA.

Recently, Gao et al.[33] designed a rapid, simple and sensitive assay based on AIE (TPBT) that can recognize dsDNA by emitting dual color fluorescent signals, red (640 nm) and green (537 nm). When TPBT binds with dsDNA, proteins, ssDNA or polyanionicanalytes it produces red color signal, while upon binding with dsDNA, TPBT specifically produced green color signal which may have been possible because of conformational changes in TPBT upon groove binding. It has also been found that TPBT can distinguish single-nucleotide polymorphisms (SNPs) in dsDNA. It could also sensitively and specifically detect DNA damage due to UV light. These methods can be a robust tool for disease diagnosis and genomics.

Recently, Zhang et al.[34] developed AIE-based strand displacement amplification (SDA) and DNA G-quadruplex detection system for patulin (PAT) toxin. This was prepared with cDNA of aptamer and PAT competed for binding to aptamer-modified magnetic beads. Primer in SDA were used to produce a large amount of G-base ssDNA in order to form G-quadruplex and obtain fluorescent signals. This has enabled to lower the detection limit to 0.42 pg mL^{-1} of PAT, which has shown to endow high sensitivity and good specificity. This study provides a reliable and stable platform for detecting small molecules. Recently, Niu et al.[35] developed AIE-based sensitive and specific fluorescence assay for DNA methyltransferase (MTase). AIE-based quaternized tetraphenyl ethene salt was prepared that binds to ssDNA through electrostatic interaction. A sequence-specific hairpin probe

Table 1 List of AIEgens used for nucleic acid detection.

Sr. No	AIEgens for DNA detection	Nucleic acids	References
1	Tetraphenylethene (TPE) based sensor	DNA	18
2	Pyridinium modified TPE	DNA	19
3	TPE with a single-stranded oligonucleotide	DNA	16
4	Alexa-azide dye	DNA	20
5	Diagonally located sulphonate groups of the tetraphenylethylene derivative (TPE-diBuS)	DNA	21
6	DNA probe (AIE–2DNA)	DNA	22
7	Positively charged fluorogen (TYPE-Z)	DNA (telomere elongation)	23
8	AIEgens modified probe (TPE-Py-DNA)	DNA	24
9	DNA-incorporated AIEgen probe (TPE-Py-DNA)	To detect DNA methylation	25
10	1,1,2,2-tetrakis[4-(2-bromo-ethoxy) phenyl] ethene (TTAPE)	DNA (to detect transgene)	26
11	Single labeled Molecular beacon (FAM-MB, with carboxyfluorescein as fluorogen and in obscene of quencher) thus simplified Molecular beacons with the aid of graphene oxide	For detection if telomerase activity	27
12	Cationic organoiridium(III) complex, 2 [PF6], with a benzimidazole-substituted 1,2,3-triazole-pyridine	Detection of ribosomal RNA	28
13	TPE N3-based sensor	miRNA for cancer detection	29
14	Ruthenium(II) complex of 4,7-Dichloro Phenanthroline	Detection of ribosomal RNA and imaging	30

was designed for DNA MTase. In presence of DNA MTase, methylation reaction started DNA polymerization with terminal deoxynucleotidyl transferase (TdT) enzymes that in turn activated fluorescence intensity via AIE with 0.16 U mL^{-1} detection limits.

3. AIE for RNA detection

Detection of DNA and RNA play a key role in genetic engineering, genomics, bioinformatics, diagnosis, forensics and many more. Conventionally, it is detected by PCR, RT-PCR and other molecular assays. Recently, the pandemic of COVID 19 have called for a rapid, sensitive and specific tool for detection at an early stage of SARS-CoV-2. The currently available antigen-based and real time PCR tools for detection of SARS-CoV-2 faces issues with sensitivity and specificity in diagnosis. Very recently, AIEgen-based nanoparticles (AIE dots) were used for detection of biomolecules, imagine and sensing.[36]

Xu et al.[37] designed and synthesized AIE-based probes for detection of DNA and RNA. It has also detected double and single stranded DNA sequences and have shown the potential to be a nucleus dye. In a study, a microchip-based AIE used sensing demonstrated quantification of microRNA. The developers of this microchip used miRNA and TPE-DNA on microwells and followed it through water evaporation. miRNA and TPE-DNA enriched in a small spot allowed higher fluorescent intensity and sensitivity.[38,39] AIE-based RNA sensing work can be explored in near future to draw the full potential of AIE in RNA biology. Even AIE is rapidly expanding for many uses and have widen the RNA biology research with better understanding of RNA.

4. Conclusion and future perspectives

AIE is a smart fluorescence material and used for several applications. AIE-based materials have opened up a new avenue for development of sensors either in liquid or solid phase. It has recently been used for sensing small molecules, proteins, DNA, RNA, environmental pollutants, toxic chemicals, localization of protein, cells and organelles such as nucleus, mitochondria, ribosome and many more. The field of AIE materials is exponentially growing and has widened the field of diagnosis and sensing. In the near future, AIE can be expanded in many other areas to harness its potential for therapeutic, biomedical and industrial applications.

Acknowledgment

The financial support from Gujarat State Biotechnology Mission (GSBTM) (Project ID: 5LY45F), Gujarat, India to G.B. and V.S. is duly acknowledged. N.G. acknowledges the Indian Council of Medical Research, Government of India for financial assistance as

Senior Research Fellowship (File No. 5/3/8/63/ITR-F/2020). R.M., G.B., N.G. and V.S. thank Indrashil University, Rajpur, Mehsana, India for providing the infrastructure facility to carry out this study.

References

1. Hong Y, Lam JW, Tang BZ. Aggregation-induced emission: phenomenon, mechanism and applications. *Chem Commun.* 2009;29:4332–4353.
2. Hong Y, Lam JWY, Tang BZ. Aggregation-induced emission. *Chem Soc Rev.* 2011; 40:5361–5388.
3. Mei J, Hong Y, Lam JWY, Qin A, Tang Y, Tang BZ. Aggregation-induced emission: the whole is more brilliant than the parts. *Adv Mater.* 2014;26:5429–5479.
4. Mei J, Leung NLC, Kwok RTK, Lam JWY, Tang BZ. Aggregation-induced emission: together we shine, united we soar! *Chem Rev.* 2015;115:11718–11940.
5. He X, Xiong LH, Zhao Z, et al. AIE-based theranostic systems for detection and killing of pathogens. *Theranostics.* 2019;9:3223–3248.
6. Ding C, Cao X, Zhang C, He T, Hua N, Xian Y. Rare earth ions enhanced near infrared fluorescence of Ag_2S quantum dots for the detection of fluoride ions in living cells. *Nanoscale.* 2017;9:14031–14038.
7. Li Y, Zhong H, Huang Y, Zhao R. Recent advances in AIEgens for metal ion biosensing and bioimaging. *Molecules.* 2019;24:4593.
8. Gohil N, Bhattacharjee G, Singh V. Expansion of the genetic code. In: Singh V, ed. *Advances in Synthetic Biology.* Singapore: Springer; 2020:237–249.
9. Wachowius F, Attwater J, Holliger P. Nucleic acids: function and potential for abiogenesis. *Q Rev Biophys.* 2017;50:e4.
10. Bhattacharjee G, Mani I, Gohil N, et al. CRISPR technology for genome editing. In: Faintuch J, Faintuch S, eds. *Precision Medicine for Investigators, Practitioners and Providers.* London: Academic Press; 2020:59–69.
11. Erickson RP. Somatic gene mutation and human disease other than cancer: an update. *Mutat Res.* 2010;705:96–106.
12. Bhattacharjee G, Khambhati K, Gohil N, Panchasara H, Patel S, Singh V. Exploring the potential of DNA fingerprinting in forensic science. In: Shukla RK, Pandya A, eds. *Introduction of Forensic Nanotechnology as Future Armour.* Nova Science Publishers; 2019:145–185.
13. Nicklas JA, Buel E. Quantification of DNA in forensic samples. *Anal Bioanal Chem.* 2003;376:1160–1167.
14. Poolsup S, Kim CY. Therapeutic applications of synthetic nucleic acid aptamers. *Curr Opin Biotechnol.* 2017;48:180–186.
15. Christensen TM, Jama M, Ponek V, et al. Design, development, validation, and use of synthetic nucleic acid controls for diagnostic purposes and application to cystic fibrosis testing. *J Mol Diagn.* 2007;9:315–319.
16. Li Y, Kwok RT, Tang BZ, Liu B. Specific nucleic acid detection based on fluorescent light-up probe from fluorogens with aggregation-induced emission characteristics. *RSC Adv.* 2013;3:10135–10138.
17. Xu JP, Song ZG, Fang Y, et al. Label-free fluorescence detection of mercury(II) and glutathione based on Hg2+-DNA complexes stimulating aggregation-induced emission of a tetraphenylethene derivative. *Analyst.* 2010;135:3002–3007.
18. Hong Y, Chen S, Leung CW, Lam JW, Tang BZ. Water-soluble tetraphenylethene derivatives as fluorescent "light-up" probes for nucleic acid detection and their applications in cell imaging. *Chem Asian J.* 2013;8:1806–1812.
19. Wang Z, Gu Y, Liu J, et al. A novel pyridinium modified tetraphenylethene: AIE-activity, mechanochromism, DNA detection and mitochondrial imaging. *J Mater Chem B.* 2018;6:1279–1285.

20. Zhao Y, Chris YY, Kwok RT, et al. Photostable AIE fluorogens for accurate and sensitive detection of S-phase DNA synthesis and cell proliferation. *J Mater Chem B*. 2015;3:4993–4996.
21. Hiremath SD, Gawas RU, Mascarenhas SC, Ganguly A, Banerjee M, Chatterjee A. A water-soluble AIE-gen for organic-solvent-free detection and wash-free imaging of Al 3 + ions and subsequent sensing of F− ions and DNA tracking. *New J Chem*. 2019;43:5219–5227.
22. Zhang R, Kwok RT, Tang BZ, Liu B. Hybridization induced fluorescence turn-on of AIEgen–oligonucleotide conjugates for specific DNA detection. *RSC Adv*. 2015;5: 28332–28337.
23. Lou X, Zhuang Y, Zuo X, et al. Real-time, quantitative lighting-up detection of telomerase in urines of bladder cancer patients by AIEgens. *Anal Chem*. 2015;87:6822–6827.
24. Min X, Xia L, Zhuang Y, et al. An AIEgens and exonuclease III aided quadratic amplification assay for detecting and cellular imaging of telomerase activity. *Sci Bull*. 2017;62:997–1003.
25. Wu J, Hu Q, Chen Q, et al. Modular DNA-incorporated aggregation-induced emission probe for sensitive detection and imaging of DNA methyltransferase. *ACS Appl Bio Mater*. 2020;3:9002–9011.
26. Jiao Z, Guo Z, Huang X, et al. On-site visual discrimination of transgenic food by water-soluble DNA-binding AIEgens. *Mater Chem Front*. 2019;3:2647–2651.
27. Ou X, Hong F, Zhang Z, et al. A highly sensitive and facile graphene oxide-based nucleic acid probe: label-free detection of telomerase activity in cancer patient's urine using AIEgens. *Biosens Bioelectron*. 2017;89:417–421.
28. Sheet SK, Sen B, Aguan K, Khatua S. A cationic organoiridium (III) complex-based AIEgen for selective light-up detection of rRNA and nucleolar staining. *Dalton Trans*. 2018;47:11477–11490.
29. Min X, Zhuang Y, Zhang Z, et al. Lab in a tube: sensitive detection of MicroRNAs in urine samples from bladder cancer patients using a single-label DNA probe with AIEgens. *ACS Appl Mater Interfaces*. 2015;7:16813–16818.
30. Sheet SK, Sen B, Patra SK, Rabha M, Aguan K, Khatua S. Aggregation-induced emission-active ruthenium (II) complex of 4, 7-dichloro phenanthroline for selective luminescent detection and ribosomal RNA imaging. *ACS Appl Mater Interfaces*. 2018;10:14356–14366.
31. Hu Y, Cao X, Guo Y, et al. An aggregation-induced emission fluorogen/DNA probe carrying an endosome escaping pass for tracking reduced thiol compounds in cells. *Anal Bioanal Chem*. 2020;412:7811–7817.
32. Kawamura K, Matsumoto A, Murashima T. Facile DNA detection based on fluorescence switching of a hydrophobic AIE dye-labeled peptide nucleic acid probe by aggregation/disaggregation. *Int J Med Nano Res*. 2015;2:011.
33. Gao Y, He Z, He X, et al. Dual-color emissive AIEgen for specific and label-free double-stranded DNA recognition and single-nucleotide polymorphisms detection. *J Am Chem Soc*. 2019;141:20097–20106.
34. Zhang M, Wang Y, Sun X, Bai J, Peng Y, Ning B. Ultrasensitive competitive detection of patulin toxin by using strand displacement amplification and DNA G-quadruplex with aggregation-induced emission. *Anal Chim Acta*. 2020;1106:161–167.
35. Niu S, Bi C, Song W. Detection of DNA methyltransferase activity using template-free DNA polymerization amplification based on aggregation-induced emission. *Anal Biochem*. 2020;590, 113532.
36. Liu Z, Meng T, Tang X, Tian R, Guan W. The promise of aggregation-induced emission luminogens for detecting COVID-19. *Front Immunol*. 2021;12, 635558.

37. Xu X, Yan S, Zhou Y, et al. A novel aggregation-induced emission fluorescent probe for nucleic acid detection and its applications in cell imaging. *Bioorg Med Chem Lett.* 2014;24:1654–1656.
38. Chen Y, Min X, Zhang X, et al. AIE-based superwettable microchips for evaporation and aggregation induced fluorescence enhancement biosensing. *Biosens Bioelectron.* 2018;111:124–130.
39. Zhao E, Lai P, Xu Y, Zhang G, Chen S. Fluorescent materials with aggregation-induced emission characteristics for array-based sensing assay. *Front Chem.* 2020;8:288.

Index

Note: Page numbers followed by "*f*" indicate figures, "*t*" indicate tables, and "*s*" indicate schemes.

A

Acetyl-CoA carboxylase (ACCase), 106–108
Adenosine triphosphates (ATP), 187
Adipose triglyceride lipase (ATGL), 106
Aggregation caused quenching (ACQ) effect, 3–5, 64, 83, 113–114, 146–147, 180, 206–208, 216
Aggregation enhanced emission (AEE), 190
Aggregation induced emission (AIE), 8
   aggregation caused quenching (ACQ), 11–12, 64
   bacterial imaging and killing (*see* Bacteria)
   intracellular temperature sensing, 13–25
   materials
      ACQ, conceptual aspects of, 3–5
      biomedical applications, active molecules for, 6–7
      conceptual aspects of, 3–5
      design principal of, 5–6
      exciton coupling model, 2
      intermolecular interactions, 2
      NIR active, molecular structure of, 5–6, 6*f*
      transition dipoles, 2
   non-radiative decay process, 12
   nucleic acids detection
      DNA detection, 221–224, 223*t*
      RNA detection, 224
   organicluminophores, 219–220
   pH sensing, 26–47
   sensors, developement of, 220
   viscosity sensing, 48–58
Aggregation-induced emission luminogens (AIEgens), 180
   applications, 63–65
   bacterial imaging and killing, 75
      bacterial viability detection, 69–70, 71*f*
      biosensing, applications in, 67–68
      chitosan, 66–67
      multiple bacterial species detection, single platform array to, 70–71
      phage-guided AIE bioconjugates for, 73–74, 73*f*
      pH dependent detection and clearance, 72–73
      photodynamic inactivation (PDI), 65–66
      toxicology, 74–75
   fluorescence strength of, 64–65
   mitochondrial detection and imaging (*see* Mitochondria, AIEgens)
   for nucleic acid detection (*see* Nucleic acids)
   reactive oxygen species (ROS) production, 64–66
   restricted intra-molecular rotations (RIR), 64
   tetraphenylethene (TPE) (*see* Tetraphenylethene (TPE))
   water solubility, 64–65
AIEgens functional nucleic acids (AFNAs) probe, 220–221
AIE-mito-TPP probe, 198*f*, 199
AIE-2Van, 70–71
Algae research, lipid specific AIEgens in, 126–137, 127*f*, 128–137*t*
Alzheimer's disease, 104
Ammonium slats, 66
Arachidonic acid, 105
ASCP probe, 196*f*, 197
AS2CP-TPA, 90–91, 91*f*

B

Bacteria
   antibiotic resistant bacterial strains, 62
   detection, 62–63
   diagnostic techniques, 62–63
   fatal infectious diseases, cause of, 62
   identification and classification of, 62–64
   imaging and killing, AIEgens, 75
      bacterial viability detection, 69–70, 71*f*
      biosensing, applications in, 67–68

Bacteria (*Continued*)
  chitosan, 66–67
  multiple bacterial species detection, single platform array to, 70–71
  phage-guided AIE bioconjugates, 73–74, 73*f*
  pH dependent detection and clearance, 72–73
  photodynamic inactivation (PDI), 65–66
  toxicology, 74–75
  theranostics, 62–63
  whole bacterium analysis, 62–63
Berberine chloride (BBRchloride), 123–125, 125*f*
Biosensing, 67–68
BODIPY 493/503 (4,4-difluoro-1,3,5,7,8-pentamethyl-4-bora-3*a*,4*a*-diaza-*s*-indacene), 111
BODIPY 505/515 (4,4-difluoro-1,3,5,7-tetramethyl-4-bora-3*a*,4*a*-diaza-*s*-indacene), 111
BODIPY sensor, 40, 41*f*
Britton Robinson (B.R) buffer solution, 32
2-(4-Bromophenyl) acetonitrile, 187–189

C

Carbonyl cyanide *m*-chlorophenylhydrazone (CCCP), 185–186
CellMask Deep Red Plasma Membrane, 91
CellMask Green Plasma Membrane biomarker, 90–91
Cell membrane disruption, 82
Cell membrane imaging, AIE materials for, 84–85, 84*f*
  detection of Cu^{2+} in live cells and, 88–92
  examples of, 86–93, 87–88*f*
  fluorescence cell membrane imaging, 92–93, 93*f*
  mechanism of, 85–86
  overview, 82–83
  traditional fluorescent sensors, 83–84
Cell-penetrating peptide (CPP), 208
CHBQ probe, 37–38, 39*f*
Chitosan, 66–67
CNGR, 208

Coherent anti-Stokes Raman scattering microscopy, 110
Colocalization experiment, 92–93, 94*f*
Confocal laser scanning microscopy (CLSM), 89, 89–90*f*
Crystallization-induced emission (CIE), 4–5
CS-Py-BC probe, 195–196, 196*f*
Cu^{2+} detection in live cells, and cell membrane imaging, 88–92

D

DBPE-DBO, AIE active polymer, 72–73
Dengue virus, 103–104
Deoxyribonucleic acid (DNA) replication, 205–206
Diacylglycerol (DAG), 106–108
2-(Diphenylmethylene) hydrazono methyl naphthalene (DPAN), 70
Direct organelle MS, 110
DNA detection, AIE for, 224
  AIE-based facile DNA detection method, 222
  AIE luminogens (AIEgens), 221–222, 223*t*
  DNA MTase, sequence-specific hairpin probe for, 222–223
  Hg^{2+}-DNA complex detection, tetraphenylethene based sensor for, 221–222
  non-emissive tetraphenylethene derivatives (TTAPE), 221–222
  patulin (PAT) toxin. DNA G-quadruplex detection system for, 222–223
  pyridinium modified TPE, 189–190
  strand displacement amplification (SDA), 222–223
  TPBT, 222
  TPE–DNAp synthesis, 221–222, 221*f*
DPMPB-FONs, 211–213

E

Eicosapentaenoic acid, 105
Elastin-like polypeptide ELP(S), 14
Electron donor-acceptor (D-A), 53–54
Electron microscopy, 110
Endoplasmic reticulum (ER), 106
Enzyme-linked immunosorbent assays (ELISA), 110

Epilepsy, 104
Escherichia coli
 detection, chitosan, 67
 micromorphological changes of, 64–65, 65f
Excited-state intramolecular proton transfer (ESIPT), 146–147, 194–195

F

Fatty acid synthase (FAS), 106–108
Fatty-ACP thioesterases (FAT), 106–108
FcPy, AIE fluorescence probe, 214–215
Fluorescence cell membrane imaging, 92–93, 93f
Fluorescence microscopy techniques, 110–111
Fluorescent molecular rotors (FMR), 48–49
Fluoride (F$^-$), 220
Fluorogenic compounds, 123
Forster-Hoffmann equation, 48, 53–54
Free fatty acids (FFA), 106–108
Free radical polymerization (FRP), 15–17

G

Gas chromatography, 110
Glycerol-3-phosphate acyltransferase (GPAT), 106–108
Gram-negative bacteria, 70–71, 75
Gram-positive bacteria, 69–71, 75
Green fluorescent protein (GFP), 123–125, 125f

H

HeLa cells, 89–90
 autophagy process in, 161–162
 cell imaging, 149
 CLSM, 209–210
 co-localization imaging experiments in, 155
 intracellular pH sensing, 161
 lysosomes of, 161–162
Hepatitis C virus, 103–104
Hexaphenylsilole (HPS), 85, 86f
Hormone-sensitive lipase (HSL), 106
Human, lipid disorders in, 104
Huntington's disease, 104

4,4′-(hydrazine-1,2-diylidene-bis(methanylylidene))-bis(3-hydroxybenzoic acid) (HDBB), 29–31
Hypochlorous acid (HClO), 165–167, 166s

I

Immunofluorescence microscopy, 110
Inorganic nanoparticles, 219–220
Interpenetrating polymer network (IPNs), 20–22
Intracellular temperature sensing, AIE molecules, 13–25
Intramolecular charge transfer (ICT), 151–153, 187

L

Laurdan, 82–83
Leukotrienes, 105
Lipid droplets (LDs) imaging, AIE luminogens for, 102–103, 102f
 biology, 106, 107f
 biosynthesis of, 106–108
 in different organisms, significance of, 103–105
  lipid disorders in human, 104
  microorganisms for human benefits, 104–105
 lipid-specific probes, 113–137
  algae research, lipid specific AIEgens in, 126–137, 127f, 128–137t
  biocompatible AIEgen from natural resources, 123–125
  lipid-specific AIEgens system with wide emission tunability, 122–123, 124–125f
  and lysosome, 118–120, 120f
  photostable lipid-specific AIEgens, 114–116, 115f
  theranostics approaches, 120–122
  two-photon lipid droplets specific AIE bioprobes, 116–118, 117f, 119f
 research, advancement and challenges in, 108–112
  lipid detection, 110–112, 112–113t
  lipidomics analysis, 110

Lipid-specific AIEgens system
  with wide emission tunability, 122–123, 124–125f
Lipodystrophy syndromes, 104
Lipoteichoic acid (LTA), 70–71, 71f
Lipoxines, 105
Liquid chromatography, 110
Living organism, 126
Lower critical solution temperature (LCST), 15–17
Luminogens, 84–86, 86f.
  See also Aggregation-induced emission luminogens (AIEgens)
Luminophores, 3–4
Lysophosphatidate acyltransferase, 106–108
Lysosome imaging, AIE materials for
  advanced diagnostic techniques, 146
  design strategy and importance of, 146–147
  detection of, 163–173
  eukaryotic cells, 145–146
  fluorescence microscopy, 146
  lysosome-specific fluorescent bioprobe, design of, 148
  viscosity tracing probes, 148–162
Lysosome tracking probes, 156–162
LysoTrackerRed (LTR) dye, 148–149

M

Malignant pleural effusion (MPE), 213
MASPB, 213, 213f
Mass spectrometry (MS), 110
Me-O-TPE-indo, 196f, 197–198
Mitochondria, AIEgens
  apoptosis and cell signaling process, 180–181
  detection, triphenylamine (TPA) derivatives for
    D-π-A framework, 193–194, 194f
    imaging and superoxide anion tracking in living cells, 192–193, 193f
  imaging
    AIE-mito-TPP probe, 198f, 199
    cationic pyridinium groups based salicyladazine fluorophore, 194–195
    CS-Py-BC, 195–196, 196f
    dual color imaging, of cell membrane, 195, 196f
    NID-TPP probe, 199–200, 200f
    nucleolus, organelle-specific emission for, 196f, 197
    polydopamine nanoparticle (PDA), photosensitizer, 196f, 197–198
    spatial–temporal aggregation-induced emission probes, 198–199, 198f
    tetraphenylethene (TPE) probe (see Tetraphenylethene (TPE))
  Mitochondrial membrane potential (MMP), 180–181
  and mouse sperm activity, AIE probe for, 187
  near-infrared (NIR) fluorescent probe, 186
  TPE-TPP based probes, 185–186
Monoglyceride lipase (MGL), 106
Motor neuron diseases, 104
Multi-fluorescence emission, 27–28

N

Nanoscale analysis, 110
Natural resources, biocompatible AIEgen from, 123–125
Neutral lipids, 102–103
NID-TPP probe, 199–200, 200f
Nile Red (9-diethylamino-5H-benzo[a]phenoxazine-5-one), 111
Non-emissive tetraphenylethene derivatives (TTAPE), 221–222
Nuclear localization signal (NLS), 208
Nuclear magnetic resonance (NMR), 110
Nucleic acids
  AIEgens functional nucleic acids (AFNAs) probe, 220–221
  detection, AIE for
    DNA detection (see DNA detection, AIE for)
    RNA detection, 224
  genetic material, mutations in, 220–221
  life processes, role in, 220–221
  synthetic nucleic acids, 220–221
Nucleus imaging, AIE materials for
  aggregated caused quenching (ACQ) effect, 206–208, 216
  AIEgens for, 208–211
    cancer cell detection and its nucleolus imaging, 211–213

nucleic acid sensing and imaging, AIE probes for, 213–215
biological processes, 205–206
cell behaviors, 205–206
deoxyribonucleic acid (DNA) replication, 205–206
fluorescence dyes, 206
fluorophores, 206
gene expression, 205–206
radioactive probes, 206
radiopharmaceuticals, 206
restriction of intramolecular rotation (RIR), 207–208

O

Optical pH sensing, 26–27
Organic-light emitting diodes (OLED's), 83
Organic luminophores molecules, 219–220
Oxydibenzene, 122–123

P

Palmitoleic acid, 105
Parkinson's disease, 104
Patulin (PAT), 222–223
Pearson's correlation coefficient, 172–173
Peptide nucleic acid (PNA), 221–222
Peptidoglycan, 69–70
Phage guided bacterial targeting and killing, 73–74, 73f
Phenylacrylonitrile, 187–189
Phenylboronic acid, 70
Phexaphenylsilol (HPS), 4–5
Phosphatidic acid phosphatase (PAP), 106–108
Photodynamic inactivation (PDI), 64–66
Photodynamic therapy (PDT), 121
  molecules for, 183–185
  photodynamic inactivation (PDI), 64–66
Photo-induced electron transfer (PET), 163
Photosensitizer (PS), 197–198
  bacteria, selective targeting, imaging, and killing of, 65–66, 66f
Photostable lipid-specific AIEgens, 114–116, 115f
pH sensing, 26–47
pH sensitive polymers, 72
Phtotothermal therapy (PTT), 197–198

p-hydroxybenzylideneimidazolidin, 150–151
PNVCL molecules, 15–17
Poly (vinyl alcohol) (PVA), 72
Polydopamine nanoparticle (PDA), 197–198
Polyhydroxyalkanoates (PHAs), 104–105
Polyhydroxybutyrate (PHB), 104–105
Polysaccharides, 69–70
Polyunsaturated fatty acids (PUFAs), 103–104
Prodan, 82–83
Prostaglandins, 105
Pyridine-tetraphenylethylene (Py-TPE), 36–37
Pyridinium, 72

R

Raman microscopy, 110
Reactive oxygen species (ROS), 64–66, 72
Red blood cells (RBCs), 74–75
Restricted intramolecular rotation (RIR), 15, 64, 146–147, 207–208
Restricted intramolecular vibration (RIV), 15
RGD, 208
Ribonucleic acid (RNA), 215
  detection, AIE for, 224
Rotaviruses, 103–104

S

Saccharomyces cerevisiae, 104–105
Salicyladazine fluorophore, 194–195
Salicylaldehyde azines (SAs), 32
Schiff base 4-N,N-dimethylaminoaniline salicylaldehyde (DAS), 43–47
Schizophrenia, 104
Silver sulfide quantum dots (Ag$_2$SQDs), 220
Single-nucleotide polymorphisms (SNPs), 222
Sterol esters (SE), 102–103
Stochastic optical reconstruction microscopic imaging (STORM), 183
Strand displacement amplification (SDA), 222–223
Synthetic nucleic acids, 220–221

T

TBB&R110@F127 nanoparticle (TRF NPs), 22–24
TCM-1, 198–199, 198f
TCM-2, 198–199, 198f
TCM-3, 198–199, 198f
TCNTP, 208
10@F127, 162
Terminal deoxynucleotidyl transferase (TdT) enzymes, 222–223
Tetrafluoro acetic acid (TFA), 36–37
Tetrahydrofuran (THF), 92, 114–116
Tetraphenylethene (TPE), 4–5, 40–42, 114–116, 122–123, 148–149, 181, 213–214, 222
   AIEgens
      bacterial imaging, 71
      E. coli and *S. aureus*, micromorphological changes of, 64–65, 65f
      TPE-star-P (DMAco-BMA-co-Gd), fluorescence emission, 68
   mitochondria imaging, AIEgens probe
      autophagy and mitophagy detection, 182, 182f
      dual functional AIE molecule, 190
      live cell imaging, chromophores for, 182f, 183, 184f
      membrane potential and mouse sperm activity, probe for, 187
      membrane potential changes, TPE-TPP based probes, 185–186
      near-infrared (NIR) fluorescent probe, 186
      photodynamic therapy, molecules for, 183–185
      pyridinium modified TPE for, 189–190
      theragnostic probe, 191–192
      tumor cells, bio-probe for, 191
      tumor tracking, 187–189
   TPE–DNAp synthesis, 221–222, 221f
Tetraphenylethene-poly(N-isopropylacrylamide) (TPE-PNIPAM), 17–20
Theranostics, 62–63
Thin-layer chromatography (TLC), 110
Time-resolved luminescence imaging (TRLI), 199–200, 200f

Toxicology, 74–75
TPE-AC, 114–116
TPE-AmAl, 114–116
TPECNPB, 121, 122f
TPEDEPy-DBz, 189–190, 190f
TPEIQ, 190, 190f
TPETH-2Zn AIEgens, 74
TPN, 190f, 191
TPNPDA-C2 probe, 195, 196f
TPPM probe, 193–194, 194f
Traditional fluorescent sensors, 83–84
Triacylglycerides (TAG), 102–108, 109f
Tricyano-methylene-pyridine (TCM), 198–199, 198f
Triphenylamine (TPA), mitochondrial detection and imaging, 122–123
   D-π-A framework, 193–194, 194f
   reactive oxygen species (ROS) process, 192
   TPA-Pyr-Br, 192–193, 193f
   TPA-Pyr-Octane, 192–193, 193f
   TPA-Pyr-Thiourea, 192–193, 193f
Triphenylphosphonium (TPP), 185–186, 192
4-(1,2,2-Triphenylvinyl) benzaldehyde, 187–189
Tromboxanes, 105
TTPM probe, 193–194, 194f
TVP-PAP AIEgens, 73–74
Twisted intramolecular charge transfer (TICT), 22–24, 146–147
Two-photon lipid droplets specific AIE bioprobes, 116–118, 117f, 119f

V

Val-pro-gly-xaa-gly (VPGXG), 14
Viscosity sensing, 48–58

W

Wax esters (WEs), 104–105

Y

Yarrowia lipolytica, 104–105

Z

Zinc (II)-dipicolylamine (ZnDPA), 66

9780323907392